Stalemate in Technology

Stalemate in Technology

Innovations Overcome the Depression

Gerhard Mensch

International Institute of Management
Science Center Berlin

Ballinger Publishing Company • Cambridge, Massachusetts
A Subsidiary of Harper & Row, Publishers, Inc.

This book is printed on recycled paper.

International Standard Book Number: 0-88410-611-X

Library of Congress Catalog Card Number: 78-16278

Printed in the United States of America

Library of Congress Cataloging in Publication Data

Mensch, Gerhard.
  Stalemate in technology.

  Translation of Das technologische Patt.
  Bibliography: p.
  1. Technological innovations. 2. Economic history—1945– 3. Patents. 4. Depressions. I. Title.
HC79.T4M4513 338.5′42 78-16278
ISBN 0-88410-611-X

To Karin

Contents

List of Figures xi

List of Tables xv

Preface xvii

Acknowledgments xix

Part I The Fundamental Process of Economic Change 1

Chapter 1
The Technological Stalemate of the Present 13

The Road to a Technological Stalemate 17
Today's Real-Economic Climate 23
Is a World Economic Crisis Inevitable? 23
  Finessing: Gaining Time 25
  The Failure of Macroeconomic Policies 28
  What Safeguard Do the International Communities Offer? 33
Summary 35

Chapter 2
The Process of Industrial Evolution 39

The Evolving Division of Labor 43
Basic Innovations—Improvement Innovations 47

The Theory of Vanishing Investment Opportunities 51
Life Cycles of Industrial Goods 52
The Diffusion of Innovations 54
Quality Competition, Pseudo-Innovations, and the Fundamental Cause of Stagflation 56
The Law of Diminishing Return Governs Improvement Innovations 62
Safety-Valve Export 63
The Stagnation Trend in GNP Growth 66
The Stagnation Trend in West Germany and the United States 69
The Changing Trend since the Industrial Revolution 71
The Metamorphosis Model in Theory and Practice 74
Summary 82

Chapter 3
Symptoms of Stagnation 85

Current Stagnations in Industrial Branches 86
Stagnations Produce Identity Crises 88
Rapid Changes Produce the Impulse Personality 90
Identity Crisis and the Need for New Identity 90
The Stagflation Epidemic and the Limits of Economic Policy 94
Concentration Causes Stagflation 96
The Liquidation Theory of Stagflation 100
Summary 111

Part II The Transfer of Knowledge 115

Chapter 4
The Ebb and Flow of Basic Innovations 119

The Process of Basic Innovation 120
The Emergence of Basic Innovations 122
Surges of Innovations 124
The Discontinuity Hypothesis 131
Do Basic Innovations Occur Randomly? 133
Summary 135

Chapter 5
The Shortage of Basic Innovations: Due to a Lack of Scientific Creativity? 137

Is There an Echo Effect in the Transfer of Knowledge? 140
Surges of Scientific Advancement 141
Checking for Echo Effects 148
Summary 150

Chapter 6
Neglect and Hastiness 153

The Paradox of Unexploited Technology 154
The Tempo of Technical Progress 158
Lead Times in the Transfer of Knowledge 163
The Speed of Transfer Processes 170
The Interplay of Stagnation and Innovation 178
Summary 180

Part III Challenges and Chances 183

Chapter 7
A Bold Projection into the Future 189

Hurdles in the Transfer of Knowledge 190
Trend Extrapolation: Looking at the Next Surge of Basic Innovations 194
Summary 201

Chapter 8
Tradition and Transition 203

Tradition Guides the Transition 204
The Seniority Principle of Innovative Change 205
A New Tradition: The Schumpeterian Paradigm 212
Summary 215

Chapter 9
The Direction of Change 217

"Hard" vs. "Soft" Transition 217
Why Is Hyperindustrialization More Likely? 223
The Technological Bias 224
The Rationalization Bias 229
The Challenges and Chances of This Stalemate in Technology 236
Summary 238

About the Author 241

List of Figures

1–1	Economic Interdependence of Western European Countries as Indicated by Coinciding Business Cycles	16
2–1	The Evolution of the Life Forms of Man as a Process of Increasing Division of Labor	44
2–2	Life Cycles for First-, Second-, and Third-Generation Computers	53
2–3	The Diffusion of Television in West Germany from 1955 to 1965 with Different Prognoses for the Following Years	55
2–4	Phases for Life Cycles of Products	65
2–5	The Diffusion of Various Products in the Economy	67
2–6	*S*-Shaped Trend in the Gross National Product and Private Investment in the United States	71
2–7	*S*-Shaped Trend in West Germany's National Product	72
2–8	The Metamorphosis Model of Industrial Evolution	73
2–9	The Metamorphosis Model of Industrial Evolution Applied to the Capital-Labor-Allocation in U.S. Manufacturing	76
2–10	Leading Sectors in German Industry 1950–76	77
3–1	The Recent Conglomerate Merger Wave	99
3–2	Merger Waves in Times of Stagnation	100
3–3	Feedback of Wage Increases on Employment	102
3–4	Michigan-Index of Consumer Sentiment in the United States	109
4–1	Swells of Basic Innovations	130

5–1 Frequency of Basic Inventions and Innovations in Chemistry and Electrotechnique Before 1900 146
5–2 Basic Inventions and Basic Innovations (First Half of Nineteenth Century) 147
5–3 Basic Inventions and Basic Innovations (First Half of Twentieth Century) 148

6–1 Acceleration in History 160
6–2 Acceleration in History 161
6–3 Lead Times from Basic Inventions in Chemistry to Basic Innovations in the Chemical Industry in the Second Half of the Nineteenth Century 164
6–4 Lead Times from Basic Inventions in Electrophysics to Basic Innovations in the Electrical Industry in the Second Half of the Nineteenth Century 165
6–5 Lead Times from Basic Inventions to Basic Innovations in the Industrial Revolution (around 1764) of Brno 166
6–6 Lead Times from Basic Inventions to Basic Innovations in the First Half of the Nineteenth Century 166
6–7 Lead Times from Basic Inventions to Basic Innovations in the Chemical and Electrotechnical Industries in the Second Half of the Nineteenth Century 167
6–8 Lead Times from Basic Inventions to Basic Innovations in the First Half of the Twentieth Century 168
6–9 Survey of Lead Times Since the Industrial Revolution 168
6–10 Lead Time and Speed of Innovative Change 172
6–11 Lead Time Decrease and Speed Increase at Various Stages in Economic History 175
6–12 The Pulsating Acceleration of Innovative Change in Economic History 176
6–13 Correlation of Frequency and Speed of Change 176
6–14 The Interplay of Stagnation and Innovation 177

7–1 Typical Inventions at the Early Age of the Patent Systems 191
7–2 Depiction of Parameters (μ, σ) of Time Series of Events in the Transfer of Knowledge 196
7–3 Parameters of Drifting Time Series of Events in the Transfer of Knowledge 198
7–4 Composite View of the Frequencies of Events in the Transfer of Knowledge 199

8–1 The Seniority Principle of Innovative Change 211

9–1 Factor Allocation in Dutch Industry 1950–1976 233
9–2 Factor Allocation in British Industry 1956–1975 233
9–3 Factor Allocation in West German Industry 1950–1977 234
9–4 Factor Allocation in Japanese Industry 1952–1975 234
9–5 Factor Allocation in U.S.-Manufacturing Industry 1950–1975 235

List of Tables

1–1 Frequency of Innovations of Different Degree of Radicalness in the Period 1953–1973 31

2–1 Kuznets Scheme of Kondratieff Cycles 39
2–2 Processes of Technological Substitution 68
2–3 Time Table of Major Crises in Economic History 78
2–4 Stagnation and the Danger of War; Peace and Danger of Economic Crisis 80

3–1 Concentration in British Manufacturing Industry, 1971 97
3–2 Stagflation and Concentration 98

4–1 Basic Innovations in the First Half of the Nineteenth Century 124
4–2 Basic Electrotechnical Innovations in the Second Half of the Nineteenth Century 125
4–3 Basic Innovations in Chemistry in the Second Half of the Nineteenth Century 126
4–4 Modern Technological Basic Innovations in the First Half of the Twentieth Century 127
4–5 Basic Innovations (Organizational and Technical) in the Industrialization of Brünn, 1740–1800 129
4–6 The Swell of Basic Innovations in the Period Around 1885–According to Different Sources 131

4–7 The Swell of Basic Innovations in the Period Around 1935—According to Different Sources 132
4–8 Test for Randomness in the Discontinuity Pattern in the Time Series of Basic Innovation between 1740 and 1960—Results 135

7–1 Events in Neopren Development 192
7–2 Sequence of Events over time in the Transfer of Knowledge 193
7–3 Time Series (simple, accumulated, and in percent) of Units in the Transfer of Knowledge 195
7–4 Frequency Distributions of Events in the Transfer of Knowledge 196

8–1 The Seniority Principle in Innovative Change during the Industrial Revolution (Brno) 206
8–2 The Seniority Principle of Innovative Change in the Nineteenth and Twentieth Centuries 212

9–1 The Rationalization Bias—Industrial Investment Targeting in West Germany, 1956–1978, According to Annual Census of Investment Plans 231
9–2 The Rationalization Bias—Industrial Innovations in Western Countries (mainly United States), 1952–1973, According to Sampling by Experts 232

Preface

The main thesis of this book (stagnation = lack of basic innovations) was conceived and developed at a time (1970–71) when economic research was still projecting incessant economic growth and when science and technology policy were expected to produce the right kind and appropriate volume of new technology to nurture this industrial evolution.

At that time, I became aware of some fundamental properties of this evolutionary process of socio-economic change. They indicated that the next downturn of the business cycle would bring disappointment to those who relied on the prognosis of "just a mild recession." Among the early indicators of a coming depression were labor market and capital market trends as well as indicators of change in the international division of labor and of technological change and substitution.

In modern industrial civilization, the boundary between nature and culture becomes increasingly fuzzy, and many of the socioeconomic forces manifest themselves in the use and creation of artifacts. Therefore, indicators on the rate and direction of industrial innovations can be expected to tell us a good deal about change within a changing economy.

The circular flow model of short-term economic activity needs a counterpart; I suggest a metamorphosis model of long-term socio-economic change. The following two propositions characterize the industrial metamorphosis:

1. Basic innovations, which establish new branches of industry, and radical improvement innovations, which rejuvenate existing

branches, tend to occur discontinuously in time, namely, in rushes.
2. Within those new or renewed branches, the pioneering innovations are followed by series of improvement innovations. The improvement effects of these successive innovations are governed, on the demand side, by the law of diminishing marginal utility and, on the supply side, by the law of diminishing marginal returns on investment.

This is classical economics. It means that in the course of time, as certain technological potentials become fully used and certain technologies increasingly perfected, there are more and more pseudo innovations that do not benefit buyers and suppliers to the extent they once did. Also, due to power, inertia, and other decisive reasons, investors tend to favor pseudo innovations over basic innovations. Consequently, reading 1 and 2 together, we can expect a lull to emerge in the course of industrial evolution—a "natural" phase in the industrial evolution. This pause in the busy creation of artifacts I call "Stalemate in Technology."

In the course of history, several such stalemated situations have occurred. They developed because of a lack of basic innovations, and ended with a spurt of basic innovations. It appears that the world economy has again become ready for another spurt of basic innovations. In the immediate future, we cannot expect the best developments at the outset but we can determine the general direction of the transition. Is the industrial world heading toward hyperindustrialization? toward a "soft" transition into some other form of post-industrial society? Judging from past experience the odds seem to be in favor of a quantum jump toward hyperindustrialization.

Berlin, November 1978 **Gerhard Mensch**

Acknowledgments

Numerous colleagues have contributed to the data and interpretations contained in this book; more than sixty referees who commented on the German version helped to improve the English one.

Special thanks go to the members of the working group on innovations research, sponsored by the Fritz Thyssen Stiftung, and to special friends: Marlene Brockmann, Ayse Kudat, Christine v. Heinz, and Werner von der Ohe.

Ms. Claire Reade translated the original manuscript, Ms. Carol Franco took out some of the Germanisms I put in when revising the first translation, and Ms. G. Hempe and H. Seiffert typed the manuscript. I want to thank them.

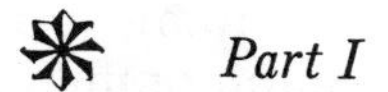

Part I

The Fundamental Process of Economic Change

"Solo in inconstantia constans"
(Benjamin Constant)

"Whether or not we admit it ourselves," Tennessee Williams wrote in *The Timeless World of a Play*, "we are all haunted by a truly awful sense of impermanence." Today two apparitions are being traded on the stock exchange for gloomy future prospects, both of which strain one's imagination to its limits. The first prophesies an impending worldwide economic crisis—who can imagine the possible depths of such a collapse? The other prophesies an incessantly creeping depression in economic life—who can say when it will end? Is there a sound basis for the widespread feeling that nothing can be done?

During the last two decades there was considerable optimism that crisis management by the new European institutions and their trans-Atlantic allies could avert a worldwide economic slowdown. A severe blow has recently been dealt to this hope, precisely because the heads of government in the most industrialized Western nations allowed their "early warnings" to publicize the fear of crisis. Instead, they should have consulted the experts who could have developed contingency plans and taken precautionary measures against the occurrence of certain critical events.

Furthermore, people no longer trust the promises of prominent economic policymakers. During the boom period of the European economic miracle and the affluent sixties, leading economists such as Clark Kerr influenced the public to believe that economic "depressions have become subject to control."[1] In addition, Geoffrey Barraclough recently mused in a report to the American Management Association that "when—in 1968, of all years!—Andrew Shonfield predicted that 'a major set-back of Western economic growth seems on balance unlikely', he was only echoing common opinion."

Evidently, a sizable remnant of this unwarranted optimism still exists among professionals and persons in positions of authority among whom, despite severe stagnation, this optimism continues to offer a pretext to avoid taking efficient preventive measures. The refusal of these individuals to face reality is understandable. For one thing, they are trapped by their earlier prognoses; moreover, taking measures to avert a crisis demands sacrifices, which they are unwilling to recommend or to implement.

In any case, the fear of crisis exists and has even affected the prosperous citizen; it has not touched only those who are in the less successful side of business. This fear has been generated in part by the widespread discussion about it, but it also has a real basis in fact. There are forces at work that Joseph A. Schumpeter characterizes as part of the "process of creative destruction." This evolutionary process of coming into existence, growing, declining, and disappearing alters all forms of economic life in the drift of structural change. Some branches flourish; others do not. Who would not rather swim along with an upward trend? Who wants to be caught in the backwash of a structural crisis in his or her industry or in the capital market?

In times of prosperity, the inevitable crisis in declining industries affects only a minority of citizens, whereas the majority lives comfortably from the effects of growth in the more modern branches of the economy. Life becomes steadily better, but this condition will not last indefinitely. There are limits to market growth determined by the amount of raw materials available, by the scarcity of living space, and by the satiation of demand for standard goods in a prosperous population. Flooded markets are a sign of this satiation, and they are the main feature of an affluent society. As growth within several of the prosperous branches of the economy approaches these limits, the crisis already occurring in the stagnant branches spreads further, thus affecting increasingly larger segments of the population. Affluence continues only in a narrowing strip of the economy.

My theory of stagnation is only a version of Say's law, which "admits the possibility of (brief) periods of disequilibrium during which the total demand for goods may fall short of the total supply, but maintains that there exist reliable equilibrating forces that must soon bring the two together." W.J. Baumol wrote in his recent article on "*Say's (at least) Eight Laws*"[2] that, given James Mill's assertion, "a nation may easily have more than enough of any one commodity, though she can never have more than enough of commodities in general,"[3] the question is how this disequilibrium can be overcome. The answer is through the provision of new types of goods and services (innovations) that better supply what people desire.

Clearly, the problems of frictional and structural unemployment of labor indicate that no reliable equilibrating forces exist that can soon bring the industrial apparatus in line with the shifting wants and needs of "affluent" people. Labor and capital have become immobile, and their reallocation in the production and distribution of alternative goods and services requires the availability of useful new technologies (recognition gap) and that needs time (reaction gap). This immobility is the temporal and technological cause of economic stagnation.

A reversal of this stagnation tendency can only be achieved through innovations that circumvent scarcities and stimulate the public's demand for new, more attractive goods and better or cheaper services. "Basic innovations" offer new chances for the blue-collar workers and the office employees who have been dismissed from their work in the stagnant branches of industry because basic innovations create new employment opportunities. These innovation projects are also the undertakings that attract capital flowing from the stagnating branches in the expectation of diminishing pofit margins.

But the ultimate question is where are these innovations? "What shall I do?" asked the rich farmer (*Luke* 12:16); "Where can I turn to?" asked the young people in *Hair*. "Can freed manpower find reemployment in the industrial sector?" is the central question in times of stagnation, today as in the past. For example, Paul Douglas asked this question in 1930 during the last worldwide slump—rather belatedly, by the way, for an expert of his stature.

The stalemate in technology is concerned with the interplay between stagnation and innovation, deadlock and progress, and crisis and revival. Refuge from crises can only be found in social and technological innovations. Therefore, the recognition that job-creating innovations may lie in the next bend in the path of stagnation will reduce the fear of crisis. Allaying this fear gives hope for the future. It mitigates the blind stupor of resignation and limits the risk of irrational outbreaks of panic, thereby confining the danger to manageable proportions. In similar situations in the past, economic depression unleashed a panic, and the business world collapsed. By reducing the fear of crisis, the potential explosion of panic is defused.

The current malaise should be seen in historical perspective with an analytical eye that borrows clairvoyance from recent advances in the theory of evolutionary processes and dissipative structures. I have been extremely fortunate in having had the chance to work closely with economic historians and members of the community of the younger generation of evolutionists. To them I owe many data and insights into the process of industrial evolution since the Industrial Revolution. The first lesson to be learned is that history, as it follows the evolutionary principles of mankind, does not necessarily not repeat itself!

There were similar situations in the past in the long run. The structural conditions of stagnation and depression, which we call stalemate in technology, have been observed in the periods around 1825, 1873, and around 1929. The world economic crises during those periods were structurally prepared for by similar chains of causation, and they erupted from overreaction and excessive fear of crisis, in other words, because those in power and control panicked.

If present conditions warrant it, we could learn from past experience in similar situations if we can perceive the analogies. We live our life looking forward, Kierkegaard said, but we comprehend it looking backward. Our understanding of various aspects of the current situation grows in proportion to the number of similar constellations we observe in the past. Data derived from an examination of Western economic history suggest that there have been several constellations that I would label stalemates in the leading technologies of the periods in question.

For such historical recurrences, the long-wave concept of phases in socioeconomic development offers a rough periodization schema that allows similar situations to be identified by when and for how long in the past analogous tendencies have prevailed. However, this intertemporal comparison is based on the premise that "history repeats itself." This is a rather ambiguous phrase, and as a premise it is, of course, disputed among economists and economic historians. We all agree that the same event cannot occur repeatedly.

But what about the recurrence of certain key similarities? Ulrich Weinstock, for example, rejects the idea that similar structural conditions in corresponding phases in industrial development can reappear, or that they could reappear at regular intervals. For the interested reader, Weinstock has collected the pertinent counterarguments and the critical literature.[4] On the other hand, Hans Rosenberg, the highly regarded historian from Berkeley, had been able to structure his vast factual knowledge only with the conceptual help of a periodization schema showing similar recurrent trends in long-term structural transformation; his survey of the supporting literature fills pages.[5]

The question is whether or not the depressions of roughly 50 years ago, 100 years ago, and 150 years ago have something in common, and if so, what is it? My answer is that the common denominator is the economic stagnation of major lines of business in the then predominant technologies that preceded every crisis. I call this phenomenon the stalemate in technology although I do not mean to imply that material factors are the overriding causes of stagnation and depression. Indeed, significant shifts in the value system lie at the root of every technological stalemate. Changes in the physical factor endowment act upon the psychological climate and vice versa. They jointly produce evolutionary transitions in human life styles.

In sifting through the literature, one's first impression is that the controversy over whether or not there has been a repetition of analogous critical constellations (about every fifty years) is simply a further symptom of a generation conflict. Of course, fathers and sons have always had quite different experiences. Clearly, only the older people could have experienced déja vue about the present constellation and the structural condition of the late 1920s. Is it perhaps because of the relative youth of today's Nobel Prize winners in economics that there has been a dearth of plausible suggestions from them about the cure of present structural imbalances that have destabilized the economy and are threatening both capital investment and the security of employment.

Conflict between generations and their dominant values is also the most plausible explanation thus far for the observed repetition of similar sociopsychological changes of climate within the course of industrial development over the last two centuries. This explanatory hypothesis puts the following observations together. What is usually called the Kondratieff cycle lasts about two generations. In addition, grandchildren think and act in ways similar to those of their grandparents. This theme runs like a red thread through the art-sociological writings of Pitrim Sorokin and the cultural-anthropological works of Alfred Kroeber, profoundly influencing their respective followers. These men studied the dialectics of attitudinal change and transformations in life-style and its interplay with socioeconomic change. These dialectics induce drastic shifts in the volume but particularly in the composition of demand for material goods and private and public services. This interplay is the forum for the forces of stagnation and innovation. One is a proponent of new, better, and cheaper goods and services and the other an opponent to such innovations. Karl Deutsch, in private communication, pointed out to me the crucial role of antiauthoritarian youth movements in this process.

I have labeled this period of reorientation "the technological stalemate." In this time of discontinuity and instability, critical developments are coming to a head, and the forces fighting for and against structural changes are clashing more fiercely. As the opposition to change must eventually give way, basic innovations are introduced that offer chances for novel investment and qualitative growth in new product and service sectors. The propensity for structural change increases sharply as antiauthoritarian twens approach forty and reach key positions.

Because structural shifts are particularly upsetting for existing large-scale operations in fully developed technologies, friction develops between the evolutionary forces of renewal and the forces understandably concerned about stabilizing the status quo by economic and political means. This results in a stop-and-go pattern of structural change. The established powers predominate in this first phase of the technological stalemate, and innovations remain only a prospect to be hoped for. In this phase, noninnovative and half-hearted attempts to overcome stagnation are initially mounted within the mature growth industries, where the slowdown of growth starts and where the difficulties in related sectors seem to originate.

As can be seen in the drug, automotive, and many other industries, the effects of minor changes are usually meager because stagnation is only an expression of the public's estrangement from a monotonous consumption pattern that is routinized by standardized supply, thus leaving little significant free choice for the user and to the public as a whole. This condition is aggravating to the individual. The public's discontent with swiftly rising prices demanded for the standardized products of the mass-producing growth industries and the powerful service industries leads to an attitude of waiting to see what will happen by the potential buyer of durable goods, and the old level of production in traditional kinds of products cannot be maintained. Even governmental policies, for example, public expenditures to boost the business cycle, cannot bring back yesterday's tastes and hungry demand. For the purpose of demand management, billions of dollars of public money were dispersed during the 1970s according to the watering-can principle, which, however, provided every investor with only a small supplement. Large-scale funding is now lacking for urgently needed innovative projects that involve small and medium-sized firms undertaking the risk of bringing better and cheaper alternatives to the market—alternatives for which the market is not satiated.

Although the explicit concern for innovation goes back at least to Daniel Defoe, who wrote an essay on it in 1690,[6] the dialectics of stagnation and innovation and, specifically, the importance of innovations for industrial prosperity and for far-reaching socioeconomic change were first brought into prominence by the great Austrian economist, Joseph A. Schumpeter. He wrote:

> "Capitalism," is by nature a form or method of economic change and not only never is but never can be stationary. And this evolutionary character of the capitalist process is not merely due to the fact that economic life goes on in a social and natural environment which changes and by its change alters the data of economic action; this fact is important and these changes (wars, revolutions, and so on) often condition industrial change, but they are not its prime movers. Nor is this evolutionary character due to a quasi-automatic increase in population and capital or to the vagaries of monetary systems of which exactly the same thing holds true. The fundamental impulse that sets and keeps the capitalist engine in motion comes from the new consumers' goods, the new methods of production or transportation, the new markets, the new forms of industrial organization that capitalist enterprise creates.[7]

Schumpeter saw that "innovations carry the business-cycle" because they are investment opportunities with additional growth potential. This growth effect of basic innovations is shown in Figure 2–10. We also want to know where the innovations originate, for—contrary to the opinion of many economists—they do not simply fall from heaven ("exogeneous variable"). As our analysis of the knowledge transfer (Part II) reveals, they emerge from the science base of new technologies and of new ways of organizing work and serving needs. The actual implementation of new technologies—that is, the timing and direction of innovation—depends upon the degree of stagnation in the old technologies and the attractiveness of the new alternatives.

Stagnation reduces the usefulness and profitability of labor and capital investments in overgrown, traditional business fields and thereby induces the implementation of cost-saving and product-adding innovations. We might add at this point that this innovation process is not restricted to any particular socioeconomic system. It is not only the market economy that is dependent on a regular stream of new technologies and reorganizations. The socialistic economies are also dependent on a high rate of growth. They prosper and evolve through the innovations they achieve, although innovativeness does not seem to have been the socialist countries' strongest point in the past.

In a technological stalemate, which is basically caused by a temporary lack of useful innovations, innovation becomes an economic bottleneck. The stalemate in technology describes a phase in the industrialization process in which many established industries approach overcapacity in the lines of standardized products and services in which they concentrate. They run into serious difficulties with these products and services on the world markets, because on one end potential consumers with enough buying power have already been satiated, while on the other end the populations at home and abroad without sufficient buying power are unable to consider purchasing these more and more expensive goods. Thus, in December 1973, only half as many West Germans ordered new cars as had done so during the same period in 1972. In an economy where 14 percent of employment depends on car sales, this was a minor catastrophy. The same sort of slump affected both British Leyland and Detroit. The oil crisis of 1973 certainly explains part of this slump, but we know now that it merely reinforced a sharp recession that was coming anyway because qualitative discrepancies between demand, and supply in many capital goods markets. This maladjustment of the supply structure is both the cause of crisis and the potential for its cure.

The technological stalemate causes a deadlock in many sectors of the economy where, consequently, particular instabilities prevail. This was the situation in the years around 1825, 1873, and 1929 when the leading Western economies were troubled by instabilities in the then leading sectors. The troubles were similar to those of today, including both structural rigidities and weaknesses. The appearance of stagnation in the key sectors, that is, in the previously most flourishing branches of the economy, were and are profoundly unsettling to the social and political positions and privileges of the beneficiaries of these branches. Stagnation plunges these groups into veritable identity crises. These beneficiaries are not only threatened with material losses, but they also face status reduction, loss of political influence, and related social costs. In terms of the wealth they have invested in stagnant industries, Marxist economists openly speak of a necessary devaluation of capital, and liberal economists and investment managers cannot and do not ignore that possibility. Inflation, therefore, may be understood as a phenomenon that shifts the burden of expected capital devaluation in stagnant lines of business onto a wider public. This is the secret behind stagflation.

Such circumstances produce sociopolitical conflict. The image of the entrepreneur, which was so highly esteemed in the past, has lost appeal in all Western societies today. Entrepreneurship is under attack. Both its achievements and its importance are called into question. This situation is very similar to the situation that occurred, for example, during the technological stalemate a hundred years ago.

Then it was not only the communists or the socialists who heaped abuse on the greedy or lazy entrepreneurs. Remarks about famous Berlin industrialists, particularly about the Siemens family, composed in rather vulgar German were heard from Wilhelm I, the Kaiser, for example. So in a way, history repeats itself.

Fortunately, a technological stalemate also affords new opportunities. A certain group of people in society pin their hopes on precisely this temporary instability in the institutional structure. The crisis in the establishment is an opportunity for evolutionary change. The lines of stagnation and innovation intersect in this temporary impasse, offering a chance for human life-style to move in entirely new directions. Revolutionaries as well as reactionaries, who always want to turn back the hands of time, sense the instabilities of the current state of affairs and try to exploit the opportunity for a doctrinaire imposition of an evolutionary course that will maximize the benefits to their group at the expense of the community.

On the other hand, reformers and innovators find it difficult to differentiate themselves clearly from these radical forces. These reformers and innovators want to promote social and technological change. They want to close the gap between supply and needs by providing better or cheaper products and services, generally without seeking to improve distribution of wealth to the exclusive benefit of any one particular group. Fortunately, the technological stalemate also gives them the best opportunity to achieve social innovations and technological basic innovations. These technological innovations venture into new industrial territory, thus giving capital and labor new lucrative tasks in areas where there is no direct competition with existing employment. We learn from history; basic innovations follow the path of least resistance and sometimes of highest need.[8]

A crisis provides the motivated and talented person with ample occasion to serve fellow human beings with useful innovations. Unfortunately, under the pressure of a crisis, there also exists the danger that useful reforms might be set aside in favor of quickly realizable technical programs, the secondary effects of which will utlimately produce a degree of harm that outweighs the immediate beneficial effects.

In times of stagnation, capital and labor that can no longer be invested in overgrown, unprofitable branches of industry provide powerful stimuli for the government to create new jobs through new large-scale technologies with military and civilian applications or both. Implementing such programs requires feasible, useful investment projects to be immediately available. However, investment projects that deviate from the old routine and promise to be profitable, namely, projects for basic innovations creating new kinds of industry or services, have never been available in large quantity. The present period is no exception. Many new technologies have been neglected for years and thereby have remained in the preliminary stages of research and development. The extent of this neglect can be imagined by looking at the old lead battery, which by 1912 was the power source for 33,000 electric cars registered in the United States. For nearly a hundred years, this power source has undergone no improvements in its energy density because research aimed at that end has offered little profit in the short run. How shortsighted, after all, is the market process of allocating investment funds to research and development activities!

Actually, the dearth of quickly realizable innovation projects is not surprising today. It has been noticeable since the mid-1960s when government programs to spur new investment failed to evoke a quick response in the business world. In particular, the severe recession of 1966–1967 dramatically brought this fact to the attention of all the experts who should have seen the problem earlier. As early as 1967, when several Western governments were resorting to expansionist policies, the unavailability of innovative investment projects was an obvious obstacle to a fast economic upswing. For example, the West German government in January 1967 allocated an extra budget for investments of 2.5 billion marks and added another 5.3 billion marks in July 1967. Altogether 11.6 billion marks of supplementary purchasing power were raised, including funds from the European Recovery Program, the new name for the old Marshall Fund. Professor Kamp reports: "When it came to the granting of these funds to suitable investors, the missing preparation for immediately realizable investment projects that would quickly produce an expansionary effect suddenly showed itself. As a result, (a) until the end of 1967, only 4 billion marks worth of public contracts found recipients, (b) only by mid-June 1968 could the total amount be paid out."[9]

By 1975 this lesson was forgotten, but it has since been learned again. In 1975 approximately nine billion German marks were being made available for expansionary programs. This amount is approximately one-fourth of the sum that President Ford promised for the same purpose in January 1975. In both Germany and the United States there was much talk of desk drawer projects that were said to be quickly realizable and only needed to be taken from the shelf. However, according to present information, these project plans dealt mostly with public investment in traditional fields such as highway construction, subsurface engineering, school building, and housing.

In short, they were in areas where additional contracts would not create any new opportunities for investment but would simply result in a higher rate of use of existing facilities.

Of course, the reason for such spending of the taxpayer's money is the hope for even larger secondary effects. According to Keynes's teaching, governmental demand management in the construction sector is supposed to stimulate further investment in other business spheres into which the publicly financed projects would supposedly spill over. In the meantime, however, the hope for such a multiplier effect from public expenditures has been largely frustrated by the lack of response in the rest of the economy. This is for obvious reasons. The multiplier effect cannot materialize if the funds are flowing into already clogged areas while the financing of basic and social innovations continues to be neglected.

The crisis cannot be cured by additional investment in already overgrown branches of the economy like the construction industry. Rather, a supplement of basic innovations in novel lines of production and service is the proper means to overcome the stalemate in traditional technologies. In September 1976, when Congress passed over President Ford's veto an act that will pump $160 million into research on electric cars, it became apparent once more that public investment allocation is no more rational than the private allocation of funds.

The predictable fiasco of policies and programs for job creation through traditional construction projects, and so on, are likely to destroy any remnant of belief in the workability of the Keynesian formula for crisis management. This would be unfortunate because this skepticism is not fully justified. Only when public investment programs involve innovative projects in which unfulfilled needs are served with useful new goods and services can an expansionary policy hope to succeed in overcoming stagnation and the dangers of crisis. Depression can be reversed only by a sufficient number of basic innovations.

We now arrive at the central thesis of this book: *basic innovations occur in clusters.* There are phases in long-term economic development in which the economy activates only few basic innovations. Then "the tiger of technical progress," as phrased by Paul A. Samuelson, falls out of step. Instead of moving onto new territory, it goes around in circles in its traditional areas. Then again, there are phases in economic development in which a swell of basic innovations passes over the stage of time. We have observed such swarms of basic innovations in periods of earlier technological stalemates, namely, in the years around 1825, around 1886, and around 1935. Each time the emergence of new innovative potentials provided new business and employment opportunities for decades and signaled the end of the respective stalemates in technology.

In the end, we give an outlook on the possibly coming swarm of basic innovations. As to the general direction in which these innovations should take effect we simply quote T.R. Malthus:

> We have seen that the powers of production, to whatever extent they may exist, are not alone sufficient to secure the creation of a proportionate degree of wealth. Something else seems to be necessary in order to call these powers fully into action; and this is, such a distribution of produce, and such an adaptation of this produce to the wants of those who are to consume it, as constantly to increase the exchangeable value of the whole mass.[10]

> General wealth, like particular portions of it, will always follow effective demand.[11]

REFERENCES

1. Clark Kerr, *Marshall, Marx, and Modern Times* (London: 1969), p. 64.
2. W.J. Baumol, "Say's (at least) Eight Laws," *Economica* (1977).
3. James Mills, *Commerce Defined* (London: 1807), p. 84.
4. Ulrich Weinstock, *Das Problem der Kondratieff-Zyklen* (Berlin: 1964).
5. Hans Rosenberg, *Große Depression und Bismarckzeit* (Berlin: 1967).
6. Daniel Defoe, *Essay Upon Projects* (London: 1697).
7. J.A. Schumpeter, *Kapitalismus, Sozialismus und Demokratie* (Bern: 1950), pp. 136, 137.
8, G. Mensch, "Basisinnovationen und Verbesserungsinnovationen," *Zeitschrift für Betriebswirtschaft*, 42 (1972), pp. 291–297.
9. M.E. Kamp, "Erfahrungen mit der Fiscal Policy," in *Vorträge aus Nationalökonomie und Finanzwissenschaft* (Bonn: 1969), p. 17.
10. Thomas Robert Malthus, *Principles of Political Economy* (London: John Murray, 1820), p. 413.
11. Ibid., p. 417.

 *Chapter 1*

The Technological Stalemate of the Present

"Tyrannical Circumstance!" (Emerson)

The stalemate in technology is a phase of upheaval in the long-term industrial evolution. It is a period of structural crisis and a time of confusion about the direction of change. "Since the future is hidden from us 'till it arrives, we have to look to the past for light on it," Toynbee wrote 1966 in "Change and Habit." The causes and effects of a stalemate in technology can best be comprehended by recalling similar occurrences in economic history. Industrial progress continues for a while despite the lull in drive and determination, but stagnation in the predominant economic areas long buoyed up by consumer demand now produces a feeling of uncertainty as to what else is socioeconomically feasible and desirable. A power vacuum and lack of momentum create a crisis of will and an environment of uncertainty conducive to both revolutionary and restorative movements as well as evolutionary changes (social and technological innovations). "Act according to your wisdom," said King David on his deathbed to his son and heir, Solomon. Unfortunately, very few governments so far have instituted reforms as wisely as Solomon did when assuming power in a time of "structural readiness for innovation."

The stalemate in technology will eventually end with the movement into another evolutionary path. Stalemated situations throughout the history of industrialization have always been surmounted with a swell of technical changes and basic innovations that establish new branches of industry and create new markets. Naturally, in a given historical situation certain new technologies will be considered more valuable than others. Now owing to the slowdown of growth in the exploitation of traditional technologies as, for example, automobile and airplane production, semiautomatic construction, chemical health restoration, and many others, the most pressing question is where to invest the accumulated capital and employ those who lost their jobs as a result of stagnation in the mature industries.

Although it is easy to see which industries stagnate, it is difficult to detect which new technologies will be the solid bases of future industries. Futuristic knowledge is usually inspecific; it is not specific enough for business leaders who would otherwise dare to venture into new investment opportunities. Of course, great value is placed on such labels as energy-conserving or environment-preserving technologies. On the same level of generality, it is easy to say that emphasis should be on new, better, or "softer," and so on, technologies that serve the needs of the population and weaken the raw material dependence of the economy. Likewise, few would disagree with the normative judgment that we need a more balanced mix of social and technological innovations. But specifically what are the projects that meet those criteria?

In this situation of depression and confusion, no government can withstand the pressure for something decisive to be done. If a government resists change, for example, on the grounds that nobody knows what should be changed and to what end, as is today the case in nuclear energy, discontent will eventually sweep it out of power. Examples of the release of such blocked pressures and social needs since the Industrial Revolution range from the storming of the Bastille and the Jacobin regime to the Nazi takeover and the Hitler regime.

What are the underlying causes that create a technological stalemate? One can examine them only with reference to the long-term developments, to the real-term substantial changes in technology on which the economy is based, and to the unfolding and shifting of wants and needs on the world markets for goods and services. The subtle interdependencies between limiting factors and opportunities can be observed, on the one hand, in stagnation and, on the other hand, in obstacles to innovation.

Today, a set of mental and material constraints holds the economic systems of the modern industrial nations in their grip. Thus, the resulting business stagnation is a world market phenomenon. The giant industries of industrialized nations face each other as competitors in both domestic and foreign markets. At the same time the increasing satiation of effective demand at home for standardized goods and services, the area of specialization for the growth industries, leaves these mammoth businesses fighting over a continually shrinking residual demand at home. The contest becomes more intense as the buying power becomes more limited because of stagnation and inflation, that is, stagflation.

National governments naturally identify themselves with their firms and their percentage of the market. With the pressure for a so-called more successful policy, they tend to shield domestic markets from foreign competition with import quotas and other productive measures, and they support domestic exporters with subsidies and export incentives.

A competitive situation in which all of the industrial countries have the same objectives and must use the same technology and policies to achieve their goals is ripe for confrontation. History teaches that it is only a short step from protectionism to outright economic war. This is the factual background of the recent disenchantment in the European Economic Community and in the Atlantic Alliance. This disenchantment evolved from open contradictions between economic policymakers' inadequate understanding of the facts combined with rapid changes in the factual basis, and this inadequacy increases the danger of the situation. Many of these structural changes have developed over a long period; and since they are of a real-economic nature, they tend to escape the attention of those who were mainly trained in monetary economics. Today these problems of structure and stability are mutual problems for most industrial countries, given these nations' close interconnections, and they have to be dealt with as common problems. The interlocking nature of the Western economies is evidenced by the close links between the movements of Western European business cycles. Figure 1–1, for example, illustrates the extent to which West German economic dynamics are linked to those of the Western European trade community.

Long-range factors, technological developments, and world market dimensions of structural change are the coordinates of the critical situation in this as in any technological stalemate. Yet what do the economic policymakers concentrate on today? The distinguishing features of their crises management policies are of a short-term monetary, economic, and chauvinistic nature. These attitudes become clear in the manipulations of interest and exchange rates in the early 1970s aimed at bettering the prospects for domestic industries abroad. This type of action caused a collapse in the Bretton Woods system; today its joints can only be mended with difficulty if at all. "Today, as in the whole of history, financial capacity and political foresight are often inversely related," Galbraith mused at the end of his study of the 1929 world economic crisis.[1]

Figure 1–1. Economic Interdependence of Western European Countries as Indicated by Coinciding Business Cycles.

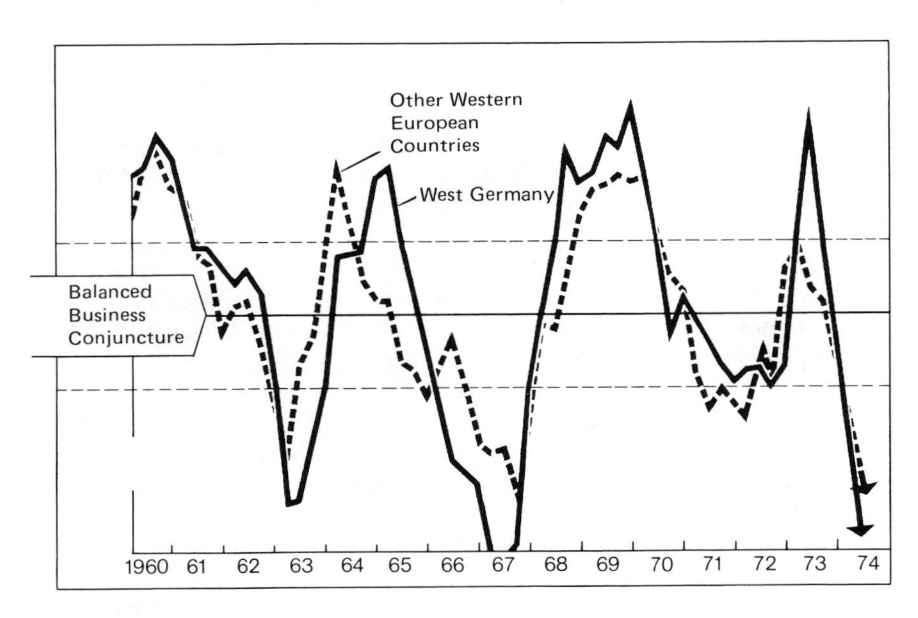

Long periods of prosperity have a destabilizing by-product; they numb the senses. A problem with the onset of the technological stalemate is that the citizens and the representatives they have voted into power cannot or do not want to consider any change in prevailing trends and therefore take no adequate precautions. Ortega y Gasset testifies about the general mood on the eve of the 1929 crisis in his book, *The Revolt of the Masses*, which he wrote in 1927 and 1928; that is, during the time of high prosperity:

> What did life appear to offer to the masses of humanity which the nineteenth century was producing in ever greater numbers? Above all, it promised general material well-being. Never had the average man been able to solve his economic problems with such facility. While large concentrations of wealth were declining in relative terms and life was becoming more difficult for the industrial worker, the economic horizon for the average man was continually widening. He became accustomed to enjoying more and more luxuries in his daily life, and his position became increasingly secure and independent of outsiders' demands. What before would have been considered a stroke of luck to be accepted with humble gratitude, was now viewed as a right to be insisted upon.
>
> Since 1900, the worker has also begun to make his life more comfortable. However, he has had to fight to achieve this goal. Society and the govern-

> ment both are marvels of organization, but they do not thoughtfully lay the good life at his feet as they did for the middle classes.
>
> Physical well-being and easy-comfort and public order, are associated with economic improvements. Life runs smoothly; upsets and dangers are unlikely to occur.[2]

Wishful thinking prevails when people shut their eyes to the dangers of crisis and deny significance to the first warnings that are given. Jean Cocteau wrote about that time:

> In 1929, people believed they had attained the pinnacle of wealth, an American apotheosis of machines, sky scrapers, science and material comfort.[3]

Similarly, Nietzsche[4] wrote: "It is discomforting and unsettling to hear talk of change"; and he said this of his contemporaries' attitudes in the dawning of the Great Depression, which came in the 1870s.

THE ROAD TO A TECHNOLOGICAL STALEMATE

We now want to sketch the sequence of events that leads to a technological stalemate. This will require skimming quickly over long stretches of history. In the 1930s, at the beginning of our forty to fifty years' journey through time, a cluster of basic innovations was affected. These innovations relaxed the grip of the deadlock in traditional technologies and ended the economic depression because they opened new fields of activity for work and capital. Even if the direct benefits of the new industries (for example, television, color films, and communications; jet engines, radar, and modern transportation; plastics, artificial fibers, and many new materials) could at first only be realized by a few, the indirect effects (demonstration that the future has many opportunities) produced a significant turnaround of the economic climate. It ended the confusion and the crisis of will that haunted the late 1920s and early 1930s.

Historically, the industrial evolution is a phase transition from one stalemate in technology to another with time spans of about two generations in between, during which diffusion and diversification of the new technologies run their course. This occurs in the following sequence:

1. In the late stage of the technological stalemate, in the years around 1935, a group of basic innovations produced several new industrial branches. (Note that the same kind of innovative spurts

occurred in the years around 1825, 1886, and 1935; each time these additional channels of activity were able to revitalize the then crippled economies.)

2. After the first factories in the new industrial branches had been built on basic innovations and the new types of goods had awakened the population's demand and enlightened investors' spirits, these branches began to expand quantatively and qualitatively through improvement innovations both in their product lines and in the production process. This set growth cycles (product life cycles or market acceptance cycles) in the new branches into motion. The movements of these cycles were closely interconnected; that is, because of the complementary relationships between goods "that simply belong together" in a modernized way of life, these growth cycles ran in parallel directions both yielding to and creating the aggregate business cycles.

The synchronization of branch cycles via complementarity relationships (cars, highway construction machinery) has not been studied extensively by economists, who by neo-Marshallian training tend to focus instead on divergent branch development caused by substitution (butter, margarine). Only occasionally has the "Veblen effect," which Thorstein Veblen wrote about in *The Theory of the Leisure Class* (1899), been dealt with by Pigou[5] and a few other scholars.[6] Hayakawa and Venieris,[7] 1977, analyzed the Veblen effect and concluded:

> Complementarity in the theory of consumer behavior is seldom mentioned except in the case of clear-cut technical complementarity (such as between bodies of cars and tires); and
>
> psychological complementarity emerges from consumption technologies associated with particular life-styles; and
>
> psychological complementarity is no less important than its technical counterpart.

In short, it is important to notice that many industries go with the same rhythm through the stages of their market cycle.

3. Parallel branch growth in new industries may create a true economic miracle throughout the country. In the United States, this was experienced during the 1940s; in other European countries it occurred only after the war when the desire of Europeans to possess or enjoy the new products and services became feasible and even became imperative in the postwar life-style. In the 1950s, one simply *had* to have the new goods (TV, computers, plastics, etc.) as they

appeared. In most markets for modern durable goods, long delivery periods indicated this seller's market. This was a period of excess demand; economic growth spurted sharply upward and was limited only by the scarcity of labor and the restricted ability to rapidly expand the production capacity. Exponential growth curves became the standard image of the further expansion.

4. During the period of expansion, the growing enterprises became dependent upon exports for an adequate profit margin and for balancing smooth demand growth and stepwise growing production capacity. Through concentration and expansion of their capacities, the competitive firms had been able to increase their efficiency and reduce their labor costs per unit. In order to maintain this profitable mass production, the firms had to sell the excess lots abroad that evenly growing domestic demand did not instantly absorb. Leapfrog capacity growth and export propensity go hand in hand.

5. The various growth industries, which because of the clustering of basic innovations, started up at about the same time and then expanded in joint leaps by the synchronization of improvement innovations eventually grew to such dimensions that their productive capacities were significantly larger than what the domestic market could absorb. This situation of excess supply meant that the large firms had to export their products whether they wanted to or not if they did not want to contract in size. Corporate growth reaches a sensitive threshold when growing productive capacities consistently exceed the increasingly satiated consumer demand. Excess supply indicates buyer's markets. It is the sign of a prosperous society since by definition the population is now amply furnished with the mass-produced goods of daily use that dominate the way of life of that period.

6. In the interim, however, similar modern and specialized industries have developed in other countries that are also approaching the threshold order of magnitude and also must rely on large-scale exporting. At this stage, however, because of standardization for the sake of mass production, the differences in the quality of technological products achieved by the various international corporations is reduced to a few outstanding features (market segmentation). Product variation now means little to the potential buyer, whom the new brand serves little better than the older one and who has little reason to acquire the newest model given that the buyer has a nearly as perfect one in his or her garage. At this stage, new vintages of the well-known investment goods and consumer goods rarely differ substantially and in fundamental characteristics from yesteryear's vin-

tage; they become interchangeable. The qualitative incentive to buy the latest product innovation diminishes to the extent that the newest offerings are merely pseudo-innovations.

Consumer decisions to buy new products then turn on unimportant distinctions, and the total volume of final demand stagnates. The large firms vie more competitively for the residual demand on the world markets. The level of demand stagnates because the unchanging supply structure diverges from the changing demand structure.

7. This threshold is also visible in the rate of return. It alerts the capital investors that it would be better not to put back into this business the huge amount of funds from net sales accumulating in their bank accounts. Rather, it seems safer and more profitable to invest this money in the international capital and money market. In 1970, every fourth mark earned by German firms was reinvested within the firm; by 1976, only every tenth mark was put back into the existing line of business where it was earned.

8. A huge money and capital market builds up. According to the figures of the World Bank, the Eurodollar market (where convertible currencies can be borrowed up to twelve months) consisted at the end of 1976 in a volume of 550 billion U.S. dollars; this was an all-time maximum in which 150 countries had a hand. The ten largest Western industrial nations are both the largest lenders (320.7 billion dollars) and the largest borrowers (271.3 billion dollars), compared to which the lendings of the Arabian oil countries (60 billion dollars) and the borrowings of the Soviet Union (10.3 billion dollars) appear small.

A few years ago, monetarists have tried to explain the growth of the Eurodollar market by the large war spendings and foreign investment of the United States. Today they are searching for an alternative explanation, a theory that could explain why this market grew by 100 billion dollars in the year 1976 alone.

I suggest the liquidation theory (Chapter 3) for explaining this surge in liquidity. It is the effect of industries' inability to reinvest all their earnings in the traditional technologies and the temporary lack of innovative investment alternatives. "Wealth may not produce civilization, but civilization produces money," Henry W. Beecher once said very much to the point here in question.

In a technological stalemate, when the industrial civilization reaches a plateau in terms of the dominant technologies and the respective way of life, financial managers in the large firms (the reference points of modern achievement and "civilization") cannot resist the movement toward lucrative currency speculation and paper investment budgeting even if they would rather not participate in such activities. The central banks of the various advanced countries, in an attempt to offset the inflationary effects of such financial investments, inject billions into the capital market system with every stability measure they are forced to take by law, using laws that are completely out of kilter with the underlying market trends. This money is credited to the speculators' accounts while it burdens the taxpayers.

9. During this stage producers from the industrialized nations congregate before potential buyers and purchasing offices in the Third World and the East. They are able to conclude agreements with these agencies, which are mostly public agencies, only by consenting to generous credit terms because the decline in prices of raw materials and already existing debts have left these countries unable to pay. Whatever the most traded goods were in an era, a technological stalemate in the international division of labor at that period occurred when the intensity of competition among the most advanced countries suddenly increased while the barter terms of trade with the least advanced nations became critically unbalanced.

These, then, are the signs of the present technological stalemate. The large industries in the industrialized nations are competing in both domestic and foreign markets for the marginal buyer; simultaneously, and contrary to all economic sense, since the North on balance is supplying both the money and the expensive finished products to the South and the East, everyone asks who will eventually assume the loss from this trade? In addition, there is a tremendous buildup of capital that is not invested substantively in productive plant and machinery but into financial investments that in the final analysis offer little more stability than a giant bank that owes more than it has. Monetary reserves, continually threatened by rumors of devaluation, become riskier assets all over the world, and the international currency situation becomes a game with paradoxical rules. One tries to acquire as much currency as possible at the same time that one attempts to get rid of it as fast as possible in order to let the threatened devaluation wreak havoc with someone else.

Stagflation is a phenomenon that many monetarists have difficulty understanding. For them, inflation and deflation are contradictions, whereas actually in times of stagflation there is an increasing supply of money the inflationary effect of which depends on the acceleration or slowdown of this money's velocity, respectively. For example, the U.S. money supply (as a percentage of GNP) was rising during the most critical years between 1924 and 1935, whereas turnover of this money decreased sharply in the years of peak liquidity, 1931–1933.

United States—Money Supply in Percent of GNP

1924	54.0%	1930	59.5%
1925	54.7%	1931	69.0%
1926	54.5%	1932	77.2%
1927	57.5%	1933	73.1%
1928	58.1%	1934	68.1%
1929	59.1%	1935	68.0%

What is the secret behind this paradoxical behavior of so-called liquidity? Surely, instead of idling around, investors would prefer this liquidity to flow to nonstagnate lines of business. As I said before, in 1970 in West Germany one-quarter of profits was reinvested in the line of business it was generated in; prior to that time, the reinvestment ratio was much higher. In 1976, this ratio had fallen to a low of one-tenth; of ten marks earned, nine went out of the business. Investors would gladly seek the safer haven of investment in profitable industrial innovations for these funds, but where are these innovations?

In a technological stalemate, return on investment in traditional technologies has decreased below the attainable capital market gain, and money floods into savings accounts, paper bonds, stocks, and other liquid assets. The reallocation of this liquidity into new, productive, safer, and more profitable investment projects is temporarily limited by a low rate of newly developed, useful, and more needed product and service innovations that would be the alternative investment opportunities.

This disequilibrium in the investment possibility mix generates structural readiness for basic innovations. In all stalemated situations since the Industrial Revolution, these preconditions for shifts in the economic structure have existed and eventually produced spurts of basic innovations (see Chapters 4 and 7 for my guess when a new spurt of basic innovations may happen in the future). Even on the eve of the Industrial Revolution, which came about through a cluster of basic innovations, and which I studied in a book with Herman Freudenberger,[8] these preconditions for radical change troubled the capital market. David Hume wrote in 1752: "No man will accept of low profits, where he can have high interest, and no man will accept of low interest, where he can have high profits."

As I go over these pages, in July 1977, the interest rate in the German capital market has dropped to the lowest level since twenty-five years as did the exchange rate of U.S. dollars. The international capital market clearly becomes ready for returning funds into less risky alternatives in business.

TODAY'S REAL-ECONOMIC CLIMATE

Given the conditions of acute stagflation (stagnation in most of what were the most powerful growth industries combined with price inflation), given the huge dollar-liquidity surplus, the germinating export trade protectionism, and the scarcity of raw materials, disquieting questions intrude. How stable is the world economy? Is the economic system fundamentally out of phase?

Yes, it has happened again. As has occurred many times in the past on the eve of the economic crises of 1825, 1873, and 1929, the economy is experiencing the constellation of circumstances that I call a stalemate in technology. This is how the situation appears today.

The world economy is completely out of phase from the North-South perspective; the differences between the wealthy industrialized nations and the economically poor, population-rich developing countries are larger than ever before. Money and goods that mutual trade should cause to flow in both directions in order for all parties to be satisfied, on balance today are mainly flowing in one direction. The monetary system is also completely out of phase. The huge dollar surplus on the Eurodollar market is being tossed from hand to hand and is under pressure to return to the United States. However, that will not work in reality.

The overarching relationship between national economic policy and structurally changing world economy is also out of phase. The policies do not harmonize with each other and with the facts. The dynamics on the world markets and the national interests in the prosperous societies directly oppose each other as the negotiation over the new economic order made unmistakably clear. A breakdown of the world economy therefore seems unavoidable to some, whereas others say "The mood is worse than the situation" (Chancellor Helmut Schmidt, for example).

IS A WORLD ECONOMIC CRISIS INEVITABLE?

To explore the question of whether a deep crisis can be avoided in a technological stalemate, it is helpful to compare a chess game with the clinch hold of the powerful, overexpanded growth industries throughout the world. Exports and import substitution is the battlefield of the economic contest in which governmental interferences lift the level of conflict from the marketplace to the political arena.

First let us compare the objectives of the game. The clear goal of the international contest on the world markets is to take customers away from foreign competition. When a situation arises in chess where neither player can win, and when no competitor in world trade can find enough customers, we face a stalemate. How can one move out of a stalemate? There are several possible ways.

The professional chess player will tend to start a completely new game with the opponent (*innovation solution*) instead of moving his or her remaining players around the board (*finessing*) in the hope that the other player will eventually make a mistake from which the first player can profit.

The professional player who wants to play the opponent again in the future will certainly avoid the folly of beginning to eliminate all the chessmen in rapid succession, tit for tat, person by person, since that would only anger his or her opponent, would not bring victory any closer, and would just increase the number of "corpses" (*pillage-chess*).

In principle, the industrial nations have these same chess moves and cooperative strategies available to overcome the technological stalemate in which their leading industries are languishing. In principle, the individual nations in this situation have serious concerns about the employment levels in the afflicted industries. They might therefore very well engage in pillage-chess. The result would be a rapid mutual exclusion of foreign traders on most markets and a collapse of world trade. This would break off cooperation without leaving any underlying desire for a new round of interaction.

This scenario, or a very similar one, has been played out every time in our economic history that a stalemate in technology has occurred, and it has always been a highly charged and explosive occasion. A few relatively (by world standards) small nations would deviate from the mutually agreed course of action, thinking that they would be able to achieve an advantage without being discovered. However, the wrath of the large nations would inevitably descend and the tragedy of pillage-chess would take its course.[a] One stimu-

[a] In the last economic crisis, Kindelberger observed that "the smaller countries, Belgium, the Netherlands, Switzerland, and Scandinavia, formed a group lacking any sense of responsibility. They converted sterling into gold and zealously raised tariffs. There is no universally accepted standard of behavior for small countries, however. Because they lack the power to affect the outcome of

lating factor has always been not to invite the smaller nations to the economic summit conferences.

A world economic crisis can only be avoided if the great powers conduct themselves in a professional manner. This requires that they refrain from pillaging and that they do not allow themselves to be led astray by the actions of the small states. The existence of a Eurodollar market poses the greatest danger in this situation. Hundreds of billions of credits are committed on a short-term basis only and are leading a vagabond's life with the potential to trigger a massive outbreak of panic. Compared to this situation the moodiness of the petrodollar is relatively unthreatening, both because of the limited volume of petrodollars involved and because they have been recognized as a political risk.

A *necessary* condition for avoiding the threatening crisis is to refrain from taking any step that might appear as a move toward pillage-chess to one's partners in the fragile monetary order and to one's competitors on the besieged world markets. Even an erroneous impression could trigger immediate retaliation. A *sufficient* condition for a short-term avoidance of the crisis is the ability to finesse. Only the innovation solution, however, offers a sufficient guarantee of long-term elimination of the danger of crisis.

Finessing: Gaining Time

Let us now scrutinize the art of finessing, that is, patiently continuing to play the game despite the awareness that the current match promises no large profits but only high risks. Today this means proceeding with export trade although the economic advantages really do not justify the labor costs. One simply has to accept the low profit game for a while in order to avoid the high cost of a spiraling crisis.

What are the real-economic parameters for finessing in a technological stalemate? The stream of goods normally flows to the places in the world where the net advantage is comparatively largest and where a technological gap exists in favor of the place of production. Normally, therefore, a division of labor exists between countries; each country produces whatever it is a position to produce, technologically and financially, on the different levels of industrial development.

The OECD study on "Gaps in Technology" found that in the mid-1960s there were still considerable gaps between the technical profi-

---

important events, they are given the privilege of following their national interests instead of having to concern themselves with the public good of stability in the world economy as a whole." [12, p. 314–15].

ciencies of various industrial nations, especially when compared to the United States. Since that time these gaps have been virtually closed in most production and consumer goods technologies. With the movement toward a technological stalemate the highly industrialized nations have come to resemble one another very closely in many merchandising characteristics, production techniques, in standards of living, and in living styles. The German, the Frenchman, and the American have become similar in their everyday activities at work and their patterns of consumption at least if one looks at the technical apparatus with which they work and live.

In the Western industrial nations, consumer items, buying patterns, the professions, and the various jobs available in agriculture and in many branches of commercial, industrial, and service industry sectors of course still reveal discernible differences. However, these differences offer no vast business opportunities because each national group preserves at least some individuality in their own version of the American way of life.

Naturally, there are regional and subcultural differences. One out of every five Swedes, but only one out of every nine Germans, went south to vacation in a sunny climate in 1971. More British than Italians read their daily newspapers because Italians prefer personal conversation. Americans tend to build more one-family homes than the Germans do because Americans like to sell their houses more frequently and move more often. The Dutch make more use of bicycles than populations in mountainous countries do; correspondingly, the Swiss are more often cowherds and the Dutch more often vegetable farmers. However, their symbols of material prosperity—cars, refrigerators, televisions, fur coats, and airplane trips—are substantially the same. Triggered by these wants, the supply gaps in most consumer goods markets in the prosperous societies have closed.

The markets for standardized goods for daily needs as well as luxuries have become highly saturated in almost all of the industrial nations. However, almost all of these countries have established indigenous factories for automobiles, refrigerators, televisions, fur coats, and so on, that are large enough not only to cover domestic needs amply but also to satisfy a part of the needs of their neighbor countries. Most of the prosperous Western societies could handle the volume of goods that are currently most in demand exclusively using their own capacities; that is, they could replenish any possible capacity deficit without great difficulty. At any rate, a far-reaching adjustment of the differences in wage levels between the European nations has virtually wiped out most price advantages formerly held by some low-wage countries in Europe and Japan.

Because of the postwar liberalization of international trade between the industrial nations, every producer has practically unhindered entrance into neighboring countries. Because of this fact and because the strong individualism in Western nations makes each person want to distinguish himself or herself from the others by his or her unique possessions, the diversified products of the various foreign enterprises find a ready market in the various countries, even when existing domestic products are qualitative equivalents. A full range of tastes can be satisfied with the wide assortment of wares available, and a complex interaction develops. For example, cars produced all over the world are eagerly bought by consumers in many different countries. Discriminating tastes and a diversified supply of goods allow overproduction in the individual nations to be disposed of on the world markets.

This international flow of differentiated but basically standardized goods may continue despite the direction of the stagnation trend, not primarily because of technological superiority of some goods, but for their lower price and because of peoples' demand for nuance or both. Thus, the exchange of goods of standard techniques will no longer be forcefully driven by the powerful tension resulting from large technical gaps. Buyers could actually buy exclusively domestic instead of foreign products, and we have heard the "Buy British!", "Buy American!", "Buy Italian!" slogans in recent years. The pampered pleasure in having a special style keeps trade going in traditional kinds of consumer goods like cottons and cars. If one asks why Japan sells fourteen times as many cars in the European Economic Community (EEC) than the EEC to Japan, it is clear that price differences hardly play a major role in all these consumer decisions to buy these standardized goods for daily use. Japanese cars are not that cheap; they are just different. The international commerce in industrial consumer goods is nourished mainly by small differences in quality; while the quality differences in the investment goods area seems rapidly to converge to some international norms and standards.

This luxury fuel propelling the international exchange of mass-produced items is naturally very unreliable. Historically, it has its own fashions and tends to disappear whenever conflicts arise between countries. Volkswagen dealers in the United States can calculate their daily turnover according to the kind of Nazi movie given yesterday on the late night show and whether yesterday's evening news included unfavorable news from Germany. The French cheese exporters fear Paris editorials because of possible adverse effects on cheese preferences abroad. Foreign trade among the Western industrial countries in a technological stalemate is as sensitive as Raquel

Welch's good name. Both can be maintained for years as long as no scandal comes up.

Like finessing in chess, continuation of trade despite shrinking profits is an important way to gain time, time in which no crisis breaks out and time to brew the appropriate antidote that at the moment is not available in sufficient quantity. In a technological stalemate, this antidote would be the implementation of basic innovations aimed at setting up new industrial branches and new areas of operation for service industries in which a large number of people eventually can expect to find new work (the innovation solution).

The Failure of Macroeconomic Policies

"Practical politics consists in ignoring facts" (Henry Adams), and recent economic policies were particularly practical in this respect.

The economic policymakers of the Keynesian school long believed that they had a formula to combat stagnation, namely, inflation. They believed that their growth and employment policies could be carried on under conditions far more favorable than those their counterparts were facing in 1929. This is partly true because automatic stabilizing mechanisms have been built into modern industrial societies, the most significant of which is a large government controlled share of the national product and a pluralistic economy in which the risk of crisis is purportedly diffused.

Doubts have arisen about the effectiveness of these built-in stabilizers. I am particularly dubious about the degree of truth in the assertion that a differentiated economy is more stable in all economic circumstances than one which is less pluralistic. Most Keynesians defend this assertion with the argument that the danger of crisis has a different basis in every branch. In response, I point out that because of underlying trends in technology many key sectors of growth in a prosperous society begin to stagnate at approximately the same time. The synchronized stagnation trend in the key industries has been underway for quite a while, and any additional crisis factors seem only free riders on this train. This is relevant, for example, for the oil boycott as well as for poor harvests. Given strong interdependencies in the technological basis for production and the demand for products, the risks of stagnation are not diffused among the economic branches and spread over time; rather, they tend to agglomerate in a general technological stalemate.[b] My argument is based upon

[b] This happened with the oil boycott of 1973. Its impact was blunted by the reduced demand for industrial oil; stagnation took the edge off the Arab oil threat, and the OPEC policy clearly has adjusted to the business recession in the West.

the clustering appearance of basic innovations (Chapter 4) and the parallel trends in the branches' cycles (Chapter 2).

Recent events give credence to my views. As I write in early 1977, three years after I wrote the above, the growth industries have all been going downhill, with short-lived fluctuations in special industries the rare exceptions. Indeed, the economic policymakers are not working under any better circumstances than their colleagues faced fifty years ago, in the late 1920s, when employment was already sluggish.

Many economic policymakers further believe that they can counteract the danger of massive layoffs with the use of government contracts, perhaps even by instituting a deficit spending policy. I have serious doubts about the efficacy of these public handout policies. This skepticism stems from the fact that the public investment capital, intended to create more jobs, must go into concrete projects in order for such a policy to be successful. The projects must promise a worthwhile investment for those undertaking them, and the jobs that are created must appear safe to the workers.

It is true that in times of high unemployment a government can make almost any kind of work credible. It can order the construction of double-decker highways, revitalize housing construction; it can even prop up the Rocky Mountains. It can also stimulate the cities to begin to build schools, hospitals, public buildings, and community centers. But are there not already two of each buildings at any conceivable place? Furthermore, a precautionary public works program instituted to prevent a crisis when crisis is not actually imminent will understandably arouse strong opposition from those who are still employed. It is relatively easy to oblige an unemployed precision mechanic or mechanical engineer to work on road construction, but it is extremely difficult to convince a fully employed skilled worker at Dow Chemical to give up his or her job, pick up a shovel, and begin to build roads for the security of his or her fellow workers at Dow; the worker will totally reject the idea. In short, structural unemployment is a difficult problem. As long as policymakers have no more innovative idea than to use the construction industry for the purpose of serving the entire labor market, they will only create a shortage of skilled construction workers but no further jobs. On the other end, there are certain limits to employment-boosting armament contracts with the Shah of Iran and other nations as well.

The main thesis of this book (whose empirical verification required substantial work by several colleagues in the field of innovations research in the United States, United Kingdom, Canada, France, and Germany) is that real-economic stagnation in a technological stalemate is so hard to fight because workable projects for substantial basic innovations, which could make new types of stable employment accessible, are not immediately at hand. The basic innovation possibilities will mostly be new technologies requiring specific knowledge. The creation and the transfer of knowledge from theory, research, and development into practical application have their own dynamics (see Part II). The preparation of alternatives has been retarded by neglect and barriers to innovation. During the period of rapid growth, the game was concentration and conglomerate competition between the large corporations. As firms would rather buy up innovative firms than developing new ventures themselves, they have usually cut out the internal search for alternatives and spend most of their research and development in defense of existing lines of business. GE, Dupont, and Monsanto managers estimated this to be true for 90 percent of these firms' research and development expenditures. On the other end, governmental research and development mostly was not consistent with the market.

Only near the end of 1974 did the German government prepare a package of expansionist measures, the so-called drawer projects. Other countries, for example, the Ford Administration, soon did the same. But what did they devise? Construction projects that would have been done anyway. I do not think this lack of innovativeness is simply ignorance but rather a temporary lack of better projects.

An international comparison of innovativeness in various industrial nations shows this phenomenon as a trend. A group of experts from the major industrial countries undertook this comparison under a grant from the United States' National Science Foundation in the spring of 1974. The findings are published in the NSF report "Science Indicators." A large sample of technological innovations introduced during the past twenty years (1953–1973) were collected and ranked according to their importance. The results are shown in Table 1–1. In the past twenty years, there have been a great many improvement innovations in established industries and service sectors, but only a very few basic innovations that were actually ventures into completely new industrial territory were found. Ask yourself how many new technologies you can think of as having created completely new types of markets or industrial branches in the last twenty years? If you find more than seven (as were found in the study), compare this number with the large number of basic innovations (listed in Chapter 4) that burst into existence in the years around 1935, 1886, and 1825.

The implication is that up to now Western industries have largely dwelled on the swell of basic innovations that came in the 1930s, and

Table 1–1. Frequency of Innovations of Different Degree of Radicalness in the Period 1953–1973.

Type of Innovation (according to degree of radicalness)	*Weighting (according to degree of radicalness)*	*Frequency (in the period 1953–1973)*
1. Basic innovations	32 to 35	7
2. Radical innovations	28 to 31	29
3. Very important improvement innovations	24 to 27	62
4. Important improvement innovations	20 to 23	145
5. Mundane improvements	16 to 19	239
6. Minor product or process differentiation with new technology	0 to 15	760
Range of weights	0 to 35	1242[a]
Total frequency		

[a] The lower classes of innovations are increasingly underestimated in quantity.

in the 1950s and 1960s very little basic innovation push developed on which we could expand in the 1970s and 1980s.

One can translate the stagnation trend that began in the Anglo-Saxon countries in about the mid-1960s and which, despite the business recession in 1966–1967, showed itself somewhat later in France, Germany, and other industrialized countries, into a simple formula:

Stagnation equals the lack of significant innovations in the last decade or so. If until very recently there have been so few basic innovations implemented, where will the necessary and convincing innovative projects come from on short notice?

An employment opportunity policy can barely be relevant to the need for stable jobs if it has no real-economic, credible, workable basis. Only demand augmenting innovation projects offers a plausible way out of the technological stalemate. As a crisis of quantitative growth, it demands qualitative change, which is a change in the supply structure oriented at the changed need structure.

The innovation solution has an effect similar to what would happen if chess players reaching a deadlock in their game stopped and began another match with their officers rearranged into positions with better potentials. Attempts to increase employment by resorting to conventional types of projects is like continuing the old game with the moves available in the deadlock situation, that is, moves with very limited potential. I see a vast potential for new industries evolving because of changing needs of Western peoples.

If there had been more basic innovations implemented in the past few years, it is possible that serious stagnation would not have plagued the Western economies. Instead, we see the British, French, and Germans ploughing valuable resources into technologies that are twenty-five years old, such as the light water reactor, the computer, and the jet engine. However, for all of the producers there is not enough market. As the search for good projects becomes frantic in an attempt to rekindle the economy's growth, the government efforts mobilized for this search soon come into contact with the owners of the large industries who no longer want their capital to remain vulnerable to the fluctuating paper investment market. Both are looking for opportunities for sustained commitment in real investment. These days, the oil and automobile industries compete with the electric power industry for good projects in nuclear energy and environmental protection. Capital is not the bottleneck. Owing to the lack of worthwhile projects for innovative change in the overexploited technology areas and the lack of opportunities for ventures into new areas during a technological stalemate, large quantities of capital are not utilized. The problem of underinvestment in existing businesses stems from the fact that during years of prosperity the preliminary work necessary to develop innovative projects has been neglected.

The Keynesians' expectation that they can pull the economy out of stagnation with government spending will be bitterly disappointed when it comes to actual fact. The spending boost must fall flat, and the desired stimulation of investment will not occur. The investors and consumers do not react to the rain of extra money that they can only, but do not want to, put into the overdeveloped production and consumption technologies. Government spending on the old type of projects will have no multiplier effect and therefore will not start an accelerated chain reaction by consumers and investors.

These factors are the structural and qualitative causes of the failure of Keynesian policies. In a technological stalemate, the owners of investment capital will hold out for new kinds of investment, that is, for novel production possibilities that cater to a hungry demand and therefore promise profit and growth. However, the development of new technologies takes years, and for basic innovations it takes even decades (see Chapters 5 and 6).

Therefore, government measures aimed at an innovation solution to the technological stalemate should not focus on quantitative demand management, which is at best expensive. For inducing a higher rate of innovative investment, a structure- and a stability-oriented research and technology policy is necessary. The least it can do is help to eliminate barriers to innovation, assuming that the policy is really oriented to the needs of the population. Unfulfilled needs really are the commercial opportunities available during a technological stalemate.

What Safeguard Do the International Communities Offer?

We have already pointed out one basic characteristic of the contemporary stalemated situation. The approach taken by policymakers of the nation-states does not harmonize with the realities of shifts in the world economy. For instance, the transition to zero growth, which the prosperous societies could perhaps afford to consider, cannot be actively organized internationally. Conversely, zero growth in a weak economy might cause a national catastrophy.

In this situation, are not the Western industrial nations' cooperations, namely, the Atlantic alliance, the European Free Trade Association, and the European Community, sufficient devices for the task of stabilizing the Western economies? No, they are necessary but not sufficient.

To clarify this point, it is helpful to examine both the legitimate and the illusory hopes that have been included in the construction of the economic blocs in the East and the West.

We begin by recognizing that the actual stabilizing mechanism in economic life is the stream of technological and social basic innovations. They provide qualitatively new avenues of growth if quantitative growth becomes too competitive. The practical application of this concept is the evolutionary principle, according to which mutations in organic life as well as innovations in human life-style are individual events, that is, singularities in the beginning and never beginning on the macrolevel. Clearly, any economic activity must obey the inherent microeconomic rules, and yet its success or failure in the system must be influenced by macroeconomic relationships. Manfried Eigen explains that "evolutionary developments derive entirely from individual events which become 'intensified' through the growth process and thereby are macroscopically 'depicted.' "[10]

Therefore, macropolicies cannot force microevents to occur. Micropolicies are required. The impact of external circumstances on the development of individual innovative activities usually is unduly strong. Macropolicies on the national or supranational level are usually a hindrance rather than a help for innovators. The interfering nature is nearly always characteristic of public policies. The constructive creativity of the macropowers is much weaker than their destructive potential. For example, it is far easier to suppress than to achieve something new with a big hand.

The help that a national or supranational economic policy could offer toward the realization of innovations is minimal. The macroperspective of global direction is too unfocused to deal effectively with microevents. The best that a large-scale economic policy can do to help innovations occur is to demonstrate an attitude of goodwill in order to create a favorable climate for innovations. Then distance should be kept by the policymakers. For this reason, I do not see how the bloc-building economic policy in the East, the West, and the South could possibly be *sufficient* to eliminate the danger of structural crisis because these policies contain nothing resembling an innovation-inducing mechanism.

My position should not be misunderstood, however. I do not mean to say that an integrated monetary and economic policy cannot make an important contribution to the *necessary* conditions for the avoidance of a world economic crisis in that it may prevent trade partners from starting an economic war (pillage-chess). The stronger the international solidarity, the smaller the probability that a national government will overreact to the pressures of stagnation and will break out of the economic community.

The nationalistic economic policy of individual states, particularly the large states, has a potential destructive power of incredible dimensions for the world economic union. Unconcerted measures taken without a consensus (such as extreme devaluations and one-sided currency controls, prohibitive customs duties, and so on) almost inevitably lead to skirmishes with neighboring countries. Economic chauvinism is sometimes profitable in the short run; but even in the short run, it is probably very costly under the conditions of technological stalemate.

At the other end, coordinated international economic policy is as effective a tool for inducing innovations in the market system as a spider web is for catching fish. The net rips under the slightest weight, as we have seen in international joint ventures to develop new technologies (computers, fast breeders, and so on). Multinational quality cartels when they are meant to distribute the potential benefits of basic innovations among various firms from different countries are highly unstable. They function in a favorable business climate, but they collapse under stress, especially if the new technology becomes ready for commercial use. Profit sharing is a different matter from development cost sharing.

Likewise, the manifestations of stagflation have practically crippled the European Economic Community organizations. As early as 1973, the late Edgar Salin diagnosed this crippling effect as follows:

> If it was still possible a year ago to speak of a crisis in the European Economic Community, apparently since then at least some aspects of the crisis have been eliminated. In light of the new circumstances we face, I do not say merely "apparently" but maintain that in fact we are in a condition which would be termed a coma if one were referring to human illnesses; a coma from which it is possible to return to full consciousness but which could also conceivably lead to a peaceful death. The following pages have been written to demonstrate that this is the condition of the EEC today, not in order to spread a feeling of hopelessness about this situation, but to state in bold print from the outset that both hope for survival and deadly peril are present.
>
> Why is this so? At the moment that these lines are being written, peace reigns on all fronts. Monetary fluctuations seem under control. Although the currency rates have not been steady for some weeks, at least they are fluctuating around a certain level, and otherwise there does not seem to be any real danger of a collapse in the wings.
>
> Despite this appearance, a careful analysis of the so-called physical and psychic state of the patient reveals that there are numerous factors which could make a swift collapse a reality, and that eliminating these dangers would require a degree of psychic and political power mobilization which can hardly be expected any more from the divided nations.[11]

It would be irrespondible of the Europeans to let their economic community break down without replacing it with other coordinating mechanisms. The EEC has already proved its usefulness even under stress when Italy and Great Britain needed help. EEC managed to prevent a withdrawal that certainly would have resulted in a round of damaging interchanges if not in a blow to trade and credit. The integration of national economic policies is a *necessary* step for the prevention of an outbreak of a world economic crisis, even if it alone is not *sufficient* to stabilize the economic process. Only the innovation solution is sufficient.

SUMMARY

The technological stalemate is a real-economic hiatus in the industrial evolution of the industrialized nations. In this situation, the world economy loses its momentum because of a temporary lack of demand augmenting basic innovations. This results in an intensified competition between the growth industries that can no longer feed their expansion from the residual demand in increasingly saturated markets, and the economy becomes destabilized when there are not enough innovative investment opportunities to counterbalance the forces of demand stagnation immediately. The trouble originates from shifting needs of the people, to which industry failed to adjust early enough with the creation of new types of goods and services.

The nation-states identify with their export industries and their domestic markets. The structural conditions of stagflation bring governments under pressure to design chauvinstic policies and lead them all too easily toward protectionism. As we have seen repeatedly in history, this condition could lead to economic war and a world economic crisis under the stress of a technological stalemate.[12] In the short run, these structural instabilities can be survived only by finessing, that is, by continuing to let trade flow even when the advantages are minimal. For in times of instability, small protective measures could have large damaging effects.

Finessing is necessary because the innovation cure, the only means to restabilizing the economic process in the Western economic setting, is not readily available. Basic innovations take time. The projects have either been stalled at the experimental stage; or they have been neglected during the economic boom when business was fully occupied pursuing other ends; or they have been effectively blocked.

Because of these barriers to innovation in alternative technologies and in more useful services, the supply of promising investment opportunities is sluggish; the rate of private investment is low; and even a multibillion dollar program of government contracts would not be very useful for creating employment opportunities as long as the necessary innovative element in the projects is lacking.

The Atlantic pact and the EEC can scarcely help the situation become better because macropolicy and global direction are concepts that are too broad to solve innovation problems where the essence of the issue is detail. However, they can prevent individual countries from becoming mavericks and in this way help to reduce the danger of the system's collapse.

The main thesis of this book says that basic innovations have predominantly occurred in the course of history in big spurts. As in every stalemate in technology the market system becomes favorably disposed for basic innovations, another swell of basic innovations is not unlikely to happen again in the foreseeable future. This scenario is presented in Chapter 7.

REFERENCES

1. J.K. Galbraith, *Der groβe Krach 1929* (Stuttgart: 1963), p. 266.
2. J. Ortega y Gasset, *Der Aufstand der Massen, Signale unserer Zeit* (Stuttgart: 1929), pp. 189–190.
3. J. Cocteau, *Essai de Critique indirect* (Munchen: 1956), p. 151.
4. F. Nietzshe, Jenseits von Gut und Böse, Aphorismus 192.
5. Pigou, "The Interdependence of Different Sources of Demand and Supply" *The Economic Journal* (1913).
6. Leibenstein, "Bandwagon, Snob and Veblen Effects in the Theory of Consumer's Demand," *Quarterly Journal of Economics* (1950); Pollack, "Habit Formation and Dynamic Demand Functions," *Journal of Political Economy* (1970); Krelle, "Dynamics of the Utility Function," in Sir John Hicks, *Anthology for Carl Menger* (Oxford: 1973).
7. Hayakawa and Venieris, "Consumer Interdependence via Reference Groups," *Journal of Political Economy* (1977).
8. Freudenberger and G. Mensch, *Von der Provinzstadt zur Industrieregion* (Brünnstudie) (Göttingen: 1975).
9. C. Kindelberger, *Die Weltwirtschaftskrise 1929–1939* (München: 1974), S. 314–315.
10. Manfried Eigen, "Selforganization of Matter and the Evolution of Biological Macromolecules," *Die Naturwissenschaften* 58 (1971), 466–523, p. 521.
11. E. Salin, "Die EWG im Koma," *Kyklos*, 26 (1973), 723–735, p. 723.
12. T.E. Burton, Crises and Depressions (Wells, 1966); E. Janeway, *The Economics of Crises* (London: 1968).

Chapter 2

The Process of Industrial Evolution

"The use of history is to give value to the present hour and its duty" (Emerson)

The symptoms of the technological stalemate, namely, the effects of stagflation, torment the citizens in the industrial nations. Stagnation and inflation force economic policy-makers into a "you are damned if you do and damned if you don't" situation. Today's state is analogous to others in our economic past. In order to identify structurally similar situations we shall make use of the model of Kondratieff cycles, which are shown in Table 2–1.

Table 2–1. Kuznets Scheme of Kondratieff Cycles.

Prosperity	*Recession*	*Depression*	*Revival*
Industrial Revolution Kondratieff, 1787-1842: Cotton Textiles, Iron, Steam Power			
1787-1800	1801-1813	1814-1827	1828-1842
Bourgeois Kondratieff, 1842-1897: Railroadization			
1843-1857	1858-1869	1870-1884-5	1886-1897
Neo-Mercantilist Kondratieff, 1897 to date: Electricity, Automobile			
1898-1911	1912-1924-5	1925-6-1939	

Source: Simon Kuznets, Economic Change, W.W. Norton & Co., N.Y.: 1953, p. 109.

The dates of the first and second Kondratieff cycle are established from the discussion for Great Britain; that of the third from the discussion for the United States. The specific dates for the three countries are presumably somewhat different, *but the differences are likely to be minor.* It should also be noted that Professor Schumpeter considers that the first Kondratieff cycle is not clearly shown in Germany. The table above was checked by Professor Schumpeter who has kindly suggested a few changes in its original version (Kuznets, p. 110).

This table is a phase model describing long-term trends in the economic climate that the Western economy has experienced and reexperienced in the last 200 years since the Industrial Revolution. These phases are prosperity, recession, depression, and then recovery. Economic historians have classified these trend periods or phases by observing how prices and production have risen and fallen. Kondratieff, Kutznets, and Schumpeter have delineated three recurrent successions of phases from 1787 to 1939; and we shall make use of their periodization schema (Table 2–1).

In collaboration with Schumpeter, Simon Kuznets undertook to specify the dates of the alternating trend phases. He received the Nobel Prize for economics partly because the use of this model permitted him to base his theoretical insights upon his wealth of historical knowledge.[1] Because recession, depression, and recovery are monetary rather than real-economic concepts, the question arises as to where we can fit the technological stalemate into the Kondratieff cycle schema. Our task addresses structural change within the framework of long-term business cycle fluctuations which, in the final analysis, are only the quantitative effects of the underlying qualitative changes.

Recession is the monetarist's expression of the monetary effects of real-economic stagnation. We may then say that a technological stalemate forms in the recessive phase of the stagnation trend. If this trend continues the economy may slide into a depression. In years of depression, the economy becomes structurally ready for radical change, and we have observed all technological stalemates to have been ended by an innovative surge. The recovery is the monetary expression of the new employment and investment possibilities achieved in the new economic areas set up by this swell of basic innovations. According to the Kuznets schema, technological stalemates must have existed in the depression phases of 1814–1827, 1870–1885, and 1925–1939. We will now try to describe the industrial evolution within this framework.

The technological stalemate is a hiatus in progress on the real-economic level. This structural instability is only an episode in the longer run even if it is a painful episode for the millions who are affected (a situation that adroit innovation and discerning policy can mitigate). In a technological stalemate, the division of labor in the world economy threatens to collapse because of shortsightedness and panicked overreaction by those in power. The danger of a depression looms large. For example, the Ifo Institute in Munich estimates that by 1980 another 70,000 businesses will disappear from the German market. Only basic innovations, which tap the large potential of hitherto unsatisfied needs, can effectively eliminate these dangers. Preparatory work on innovations, however, has either been neglected or delayed during the years of prosperity. These sins of omission have retarded the normal development of the Schumpeterian process of creative destruction, thereby introducing the eventuality of incurring larger social costs than are actually necessary if the adjustment were to proceed smoothly. It is imperative to prevent this eventuality from occurring, and it is therefore necessary to understand the interplay of stagnation and innovation.

In the industrial economy as well as in the natural ecology, creation and destruction are symbiotic opposites. How are they related to each other? They are both phenomena associated with industrial evolution. Modern scientists can explain the creation of life as a chain reaction in the inanimate world, that is, as a self-directed multiplication of molecular forms and a selection of those forms best suited to certain environmental conditions. The same logic also explains the development of mutants of new animal life forms according to particular evolutionary principles. Why should the same logic not apply to the development of basic innovations of new human ways of life?

The evolution of human life and work styles can then be viewed as an alternation between creation, growth, decline, and disappearance or between stagnation and innovation in the different parts of a socioeconomic system. All of these processes are part of a self-regulating cycle that historically has caused the industrialization process to progress in developmental leaps. Economic historians have perceived these processes as Kondratieff cycles; we perceive of them as spurts of basic innovations and of subsequent sequences of improvement innovations that eventually run out of steam.

One can also observe the alternation of stagnation and innovation in the timeless realm of culture by examining Western mythology. Prometheus, the fire-giver, implemented the first technological basic innovation. In Genesis, Moses outlined the dialectic of stagnation in paradise and the first social innovation that followed from stagnation. What befell the perfect arrangement in the garden of Eden is a preeminent object lesson. Paradise was an agriculturally based market economy that used a gold standard:

> "There is gold; And the gold of that land is good."[2]
>
> "And the Lord God took the man, and put him into the garden of Eden to dress and to keep it."[3]

Clearly, the first manmade innovation is given as the reason for human kind's expulsion from paradise; it was a very important social innovation. Human beings used the fruit from the tree of knowledge for their own enlightenment. The significance of this achievement is not diminished by the serpent's master-minding role. After all, the serpent is identical with the archangel who had occupied the number two position before his fall. The archangel's fall, more so his earlier intrigues, reflect both a power vacuum in the high echelons of the heavenly host and a stagnation of the traditional influence of the established executive system in paradise. This stagnation was clearly a precondition for the first innovation that occurred.

The spectrum of consequences resulting from this first innovation is also instructive; the expulsion of humankind affected both the governor and the governed. Moses tells of the blow to the so-called alliance, the heavenly host's social contract with humankind. As a result, both sides were compelled to take subsequent actions:

The heavenly host had to modify its unconditional dominance.

The population had to modify its living conditions through further innovations ("they knew that they were naked; and they sewed fig leaves together, and made themselves aprons.")[4]

Of course, one has to distinguish between stagnation of power and stagnation of performance. The story of creation and the tale of Prometheus foster no illusions as to what punishment those in power will inflict upon innovators whose grasp for power fails. They will be harrassed, symbolically trampled into the ground as the serpent was, or chained by the heels and left to the vultures as outcasts.

Human society is an innovative society. Both the harrassment of revolutionaries and the exclusion of social reformers by those in power affects an evolutionary principle as vital as that affected by the blocking or neglect of technological innovations in the industrial economy. Just as an insufficient rate of innovation in the economy eventually leads to stagnation in performance and to political stagnation, hostility to social reforms can deepen the economic stagnation, thus causing the benefits that flow from government to decline and positions of power to become problematic. A credibility gap ensues, which increases the economy's need for more flexibility and better solutions to its problems, that is, to a need for technological and social innovations. Of course, an attitude of no experimentation is attached to every form of government that must maintain itself if it is going to survive. This problem is least acute in a democratic system. On the other hand, the hostility to reforms and opposition to innovations exhibited by nondemocratic powers is understandable from this point of view, because the reforms and innovations have aftereffects that often disturb the established order. Change cannot be held within certain bounds; it metastasizes. To prevent gradual change, however, only makes inflexible structures more fragile.

The occurrence of change is therefore an inherent characteristic of societal and economic life. Modern times cannot be distinguished from ancient times in this regard. In today's modern industrial societies, it is therefore not surprising that certain technological basic innovations meet rigid resistance from many circles in society. Georgescu-Roegen, who applied the concept of entropy creation to understanding economic change, said that we would be utterly mistaken to believe that technological innovations modify supply alone. The impact of a technological innovation upon the economic process consists of both an industrial rearrangement and a consumer's reorientation, often also of a structural change in society.[5]

Schumpeter emphasized this point of view: "The capitalist economy is in essence the framework of a process of not merely economic but also social change."[6] "Without this change, or more precisely, without this type of change which we have labelled development, the capitalistic society cannot exist, because the economic functions, and together with these functions, the economic foundations of the leading classes—the classes which serve the capitalist apparatus—would collapse, if this change ceased: without innovation there are no entrepreneurs, without entrepreneurship there are no capitalist profits and no capitalist momentum. Naturally this is a highly generalized characterization of the situation. The atmosphere of industrial revolution—of progress—is the only atmosphere in which capitalism can survive."[7]

That is the simplest and at the same time the most ingenious secret of evolution—no way of life is without change. We may not easily admit it to ourself for the sake of feeling secure, but we cannot ignore it as a fact that Moses and Herodotus have written already in ancient times into the axiomatic basis of our Greeco-Judean culture.

THE EVOLVING DIVISION OF LABOR

The innovative society evolves through a Schumpeterian process of creative destruction. This is the normal process for the development and spreading of the human way of life. How does this process take place and how does it fit into the observed historical phases as outlined in the Kuznets schema above?

Our evolutionary model is an organic model that does not claim much originality as it harks back to the imaginative world of the ancient thinkers. It also fits neatly into the Darwinian concept of the evolution of species. The evolution of the human species and ways of life can be diagrammed in the form of a tree that is continually branching out throughout the history of man. As a model of differential development, it schematizes the unfolding of ways of life from prehistoric times to the present (Figure 2–1). From an economic per-

Figure 2–1. The Evolution of the Life Forms of Man as a Process of Increasing Division of Labor.

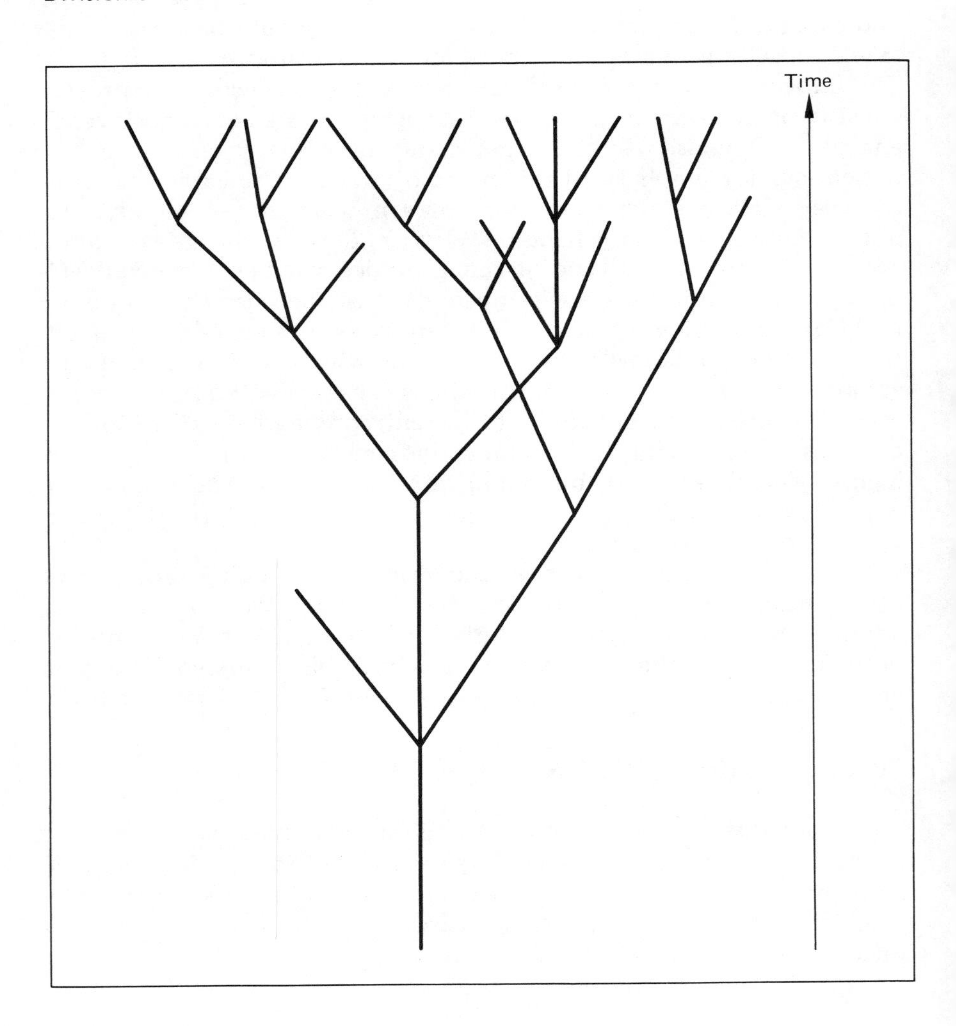

spective, one can view this genetic development as an evolutionary process of a deepening division of labor and increasing specialization.

The division of human resources into special skill areas has occurred most strikingly in Western development; the West owes its leading role in industrialization to its inseparable counterpart, urbanization.[8] Industrial and urban dynamism are two aspects of the same development;[9] they give each other momentum by mutual reinforcement. New problems create new solutions, and they produce new problems again and from these come more new forms of activity. Moreover, life in towns seems particularly conducive to an acceleration in the further deepening of the division of labor through innovation. As my work with H. Freudenberger on the innovative surge experienced during the Industrial Revolution shows, even innovations in agriculture frequently originate in the cities.[10]

Innovations expand the spread of the evolutionary tree as new forms of activity branch off. The division of labor becomes more complex more quickly in an innovative economy. As traditional obsolete professions and kinds of work stagnate and even disappear, the city grows and prospers. Structural changes in the pattern of living and working, stagnation, and innovation go hand in hand. History has always proceeded in this manner. What Herodotus wrote more than 2,000 years ago, "I wish to tell the story of the small cities as well as of the great, since most of those which were once great are now small, and those which came to greatness during my age were small enough in olden times," summarizes at least 3,000 years of human experience.

For the individual, the increasing division of labor compels specialization in the working world and identification with different roles in daily life. We will not concern ourselves here with the fact that the multiplication of roles for modern man creates a specific identity problem and a need to switch this identification too often (see Chapter 3). Here it is more significant to note that the modern person's role fragmentation has a counterpart in the individualization of consumer options for humankind. The number of one-family households is larger than ever, and the individual car transportation creates the traffic jams in modern cities. Many people have noticed that the source of the volatile nature of the market economy lies in too deep a division of labor in production and consumption. They become increasingly discontent with isolated goods consumption and alienated goods production. They long for alternatives but, of course, do not know what they might consist in. Ivan Illich speaks of a new type of hunger that is befalling affluent societies.

The fragmentation in working and daily life does not bring only a painful superficiality to human relations. It is also the price to be paid for a higher degree of efficiency in specialized and predominantly technical jobs, on the one hand, and the customs duty paid for a wider selection of available consumer goods, which are mostly technical products, on the other hand. In addition, the division of labor creates a functional separation in the exchange of goods; the workers do not know for whom they are making the products, and the consumers do not know who put their purchase together. The market as an anonymous intermediary separates more than it unifies, and often it takes away more than it gives if all factors are evaluated in a given situation. In a technological stalemate, when higher prices tend to be asked for standardized products, consumption tends to stagnate because of social and economic factors.

Since the Dukes of Venice 500 years ago instituted a system of patents that allowed the inventors of new techniques to have a quasi-monopoly on the benefits of their inventions, the deeply entrenched division of labor in the modern world has benefited technical improvements and harmed the improvement of service delivery. Among the industrialized nations, Germany and Japan's industrial sectors today make up the largest proportion of their gross national products (45 percent and 37 percent) while the proportion taken by their service sectors is the smallest (27 percent and 24 percent) according to the *United Nations Statistical Yearbook* in 1973. I think the patent system has been the main cause of or at least partly responsible for the pronounced technical bias in economic development throughout the last four centuries.

The particular form of the division of labor that we call the modern industrial society only exists for a minute in the time span of life on earth, and it will certainly branch out further with the introduction of additional innovations. The technological stalemate today again poses the question as to what is the degree and direction of desirable progress.

There can never be a completely satisfactory answer to this question; even modest partial answers require vision and information that is difficult to obtain. First, it is difficult even to orient oneself within the confusingly complex structure of the economy because the information content of the modern economic structure is much higher than that of yesterday's less fragmented economic structure. The enrichment of the information content with the progressive multiplication of specialties is an essential trait of evolution, as basic to the inanimate world as it is to the animate world. This is common knowledge to scientists. In the social sciences and humanities, questions about relevant information, its sources, and the mechanisms for their disguises have not been studied extensively. In our investigation of the essential characteristics of industrial evolution, where we face the problem of scarce data, we shall allow ourselves to be guided by the thoughts used in modern scientific analysis to make the scarce data indicate patterns.

In his chapter on the principles of selection and evolution, Manfred Eigen writes:

> The larger the content of information, the more there is a justification for separating the two processes:
>
> 1. Selection among populated alternative states.
> 2. Evolution of the selected states.
>
> Both processes merge into one if the structural capacity is so low that all possible alternative states are populated. However, the number of possible complexities is usually tremendously large.[11]

It is precisely these two processes that we want carefully to separate in the following discussion of types of innovation.

BASIC INNOVATIONS—IMPROVEMENT INNOVATIONS

The evolutionary tree of human life-styles (Figure 2–1) has numerous offshooting branches. Every fork in a branch stands for the opening of a new path—a new method of operation or a new technology, that is, a novel area of activity that can potentially offer employment to a large group of people. These forks, deviating from former practice, result from what I call *basic innovations.* Technological basic innovations produce new markets and industrial branches whereas nontechnical basic innovations open new realms of activity in the cultural sphere, in public administration, and in social services. Basic innovations create a new type of human activity. Extensive lists of examples are given in Chapter 4.

Further development in established areas of activity (which were once established by a basic innovation) is what I call *improvement innovations.* Improvement innovations account for the linear extensions of the branches on the evolution tree (Figure 2–1).

For example, one improvement innovation in the industrial sphere would be the introduction of a new product that is superior to its ancestor in its quality, reliability, ease of use, environmental protection, raw material use, labor cost, and so on. It could also be the application of new and better production techniques that would allow old or new products to be made more reliable, of better quality or simply in larger quantities, or at a lower price.

In 1971 when I first published an article on the distinction between basic and improvement innovations, it produced some professional controversy.[12] The discussion seems to have confirmed that I had made a meaningful distinction with this pair of definitions.[13] The distinction is significant for the examination of the conditions necessary for further evolution at a high level of economic development. Given the high information content in this kind of society, it is important to separate the choice among and the opening of alternative lines of development (basic innovations) from steps forward in lines of development that have already been established (improvement innovations). These two forms of innovation are explained by completely different theories and do not behave alike in practice.

In fact, our analysis of socioeconomic change and knowledge transfer in Chapters 4 to 6 shows that these two forms of innovative activities compete with each other, largely because of inertia and the problem of scarce investment funds. Research, the development of new ideas, and the practical implementation of new concepts all require labor and capital. All innovators compete for the limited research and development and venture capital funds regardless of whether they are planning to implement a basic or an improvement innovation. Moreover, completely new kinds of projects have many more problems attracting investment capital than the more usual kinds do, especially since most research and development in industry is spent in defence of existing lines of business.

But the picture is constantly changing for the innovators in this competition. At one time a basic innovation and at another time an improvement innovation will have a better chance to attract the necessary capital and labor. In industry, there are alternating priorities among engineering tasks. This is a phenomenon with which industry is very familiar although finance managers sometimes do not understand it particularly well.[14]

Priorities fluctuate according to the economic climate. In the market economy, the emphasis is almost always placed upon improvement innovations. However, the economic situation tends to become critical if the impetus from prior basic innovations peters out and more pseudo-innovations are substituted for real improvement innovations. Then in a stalemate in technology, the propensity for basic innovations increases.

It appears that the socialist planned economy generally has tended to emphasize nontechnical basic innovations[15] rather than technical improvements,[16] at least up until now. Clearly, and understandably, the policymakers in the Eastern bloc have no desire to further intensify the already pent-up demand for goods with any improvements in their quality. Providing the population with a generously varied choice of goods has traditionally been the strength of the private economy in a market system. In making such a comparison now, however, one must also take into account that the Eastern European countries once have had to start from situations of much greater privation. They therefore have given investment priority to the establishment of heavy industries and social services with little attention paid to qualitative improvements in consumer goods. And for the purpose of quickly employing as many people as possible, they often resorted to less effective, labor-intensive production techniques that in the market economy would not withstand the pressures from competition.

A basic feature of the market economy system is the greater tendency for entrepreneurs to attract potential buyers with qualitative improvements in their products. Competition feeds this tendency—at least in those branches where it has not been eliminated by regulation, for example. Antitrust measures are directed mostly against shady practices in pricing competition; they often overlook the fact, however, that many of these practices are only feasible if the qualitative competition has previously been stifled (either by a market power, barriers to entry, or by the presence of pseudo-innovations). In this situation, the antitrust laws become virtually ineffective.

Normally, the tendency of the market system is to use improvement innovations, and that keeps the display cases filled with diversified products. This has had a strong impact on the Eastern countries. It whets the desires of the East Europeans for Western-style consumer goods. These wishes are difficult to fulfill, however, if one is continually striving for innovation in the arms industry. The policymakers in the East are torn between the wishes of the civilian population and the concerns of the military, thus making the successes of both Western civilian and Western arms industries doubly painful for them.

However, positions could conceivably be reversed. It is possible that in the future people's attitudes about desirable avenues of innovation may change in favor of nontechnical developments. Many believe in a trend toward more public goods. This is really the challenge to the market economy.

In practice, Eastern governments have not availed themselves of the principal chance offered by a central planning system; that is, to accommodate their citizens with new and more attractive governmental goods and services through a higher rate of basic social innovations. The Eastern achievements could glitter as tantalizingly in the West as the Western consumer successes do in the East; in fact in the area of public health they already do.

As we have said, Western market economies generally favor improvement and pseudo-innovations over social innovations and technological basic innovations. As the technological stalemate becomes entrenched in the industrialized nations, this general trend will temporarily change. The established branches of growth industries have expanded to such an extent and the degree of saturation with durable goods is so high that investment in these branches is not very profitable. For example, the size of the European synthetic fiber industry is estimated to be 25 percent above market absorption. In such lines of business almost all of the engineers' suggestions for increased efficiency (if they also involve capacity expansion) fall on deaf ears, and product designers find it difficult to interest people in their new versions of the old products. In this situation, capital owners seek investment opportunities in new areas where the market is still hungry and there is little competition. This means that basic innovations are needed. In a technological stalemate, basic innovations are of immediate interest even to a market economy. In this situation, and only in this situation, the capitalist economy tends to pursue radically new lines of development.

The family tree of socioeconomic development (Figure 2–1) makes this graphically clear by showing that most of the offshoots occur in a few short periods. This is the phenomenon of the surge of basic innovation. The results of our empirical studies, set out in Chapters 4 through 6, show that this kind of clustering in the appearance of basic innovations has occurred periodically during the history of industrialization; that is, in the years around 1825, around 1886, and around 1935.

Because the major portion of this book will be concerned exclusively with basic innovations, the remainder of this chapter will be devoted mainly to improvement innovations and pseudo-innovations and their effects. Stagnation of certain branches of industry can be viewed as stagnation in useful technical progress (in the form of fewer improvement innovations and, consequently, more pseudo-innovations). And we conclude that this stalemate in the overutilized technologies first creates the structural readiness for basic innovation in the entire economy (macrolevel) and then the willingness of individuals (microlevel) to try new ways of doing things, trying basic innovations, investing capital and labor into them, and making the best of the chances they offer to him who takes the risks of venturing into unfamiliar industrial territory.

THE THEORY OF VANISHING INVESTMENT OPPORTUNITIES

When in the 1930s the need for a sound theory of stagnation was held by all economists, Schumpeter did not accept the theory of diminishing investment opportunities that had been derived by Alvin Hansen from Marx's theory of capital. This type of stagnation theory sees the world economic crisis as a collapse of production and trade whose advent is foreordained by the general slackening of investment activity in the capitalist countries. The sluggish rate of investment as a result of lacking business opportunities leads to structural crisis. It leads to system change. As the government steps in with public investment, it crowds out private investment, and the theory predicts an eventual transformation of the economic order.

This theory is partly true and partly false. Schumpeter did not accept its premises (general lack of investment opportunities) nor its conclusion (revolution). He could not imagine that human inventiveness and motivation could be completely exhausted. I cannot either. The true part has been diagnosed by Keynes: "The weakness of the inducement to invest has been at all times the key to the economic problems."[17]

Schumpeter believed that capitalism surely would die out, not because of a lack of opportunities or a weakness of the inducement to invest, but because capitalism on the whole would suffocate from its obvious expansionary successes. Schumpeter thought these successes would alienate the social groups who safeguard, maintain, and encourage the market processes by arousing a countermorality that would undermine the capitalist institutions themselves and point to socialism as the only conceivable successor to the capitalist system.[18]

Like the Marxists, Schumpeter applied the theory of the vanishing investment opportunities to the market system as an entity. By rejecting this theory completely, he managed to throw the baby out with the bathwater. Schumpeter (and many neo-Schumpeterian economists, for that matter) did not differentiate between industrial sectors and types of innovation. They therefore did not adequately consider the fact that this theory of dwindling investment opportunities provided the best explanation for the circumstances of structural instability and of partial growth in certain special areas (structural change on the micro- and mesolevel that would in effect stabilize the macrosystem).

The valid aspect of the theory of vanishing investment opportunities is known as the product life cycle theory. It explains the change in frequency and utility of successive improvement innovations in individual industries that have a particular technological base. It also provides a sequencing model for successive product and process innovations in certain industrial branches. In this way it depicts how a basic innovation unfolds and how it is followed by many genuine improvements during the times of this industry's prosperity, which in turn eventually gives way to a rising flood of pseudo-innovations in its stagnation phase. Stagnation in industry is primarily a mesoeconomic phenomenon; it is only when mesoeconomic stagnation appears in several branches at the same time that it becomes visible on the macroeconomic level and in society at large. In a technological stalemate, the trouble is that a number of branches reach their stagnation phase simultaneously.

Life Cycles of Industrial Goods

The theory of vanishing investment opportunities is far from uncontroversial among macroeconomists. On the other hand, it poses no problem to the microeconomist if he is familiar with the modern theory of corporate growth. The theory of corporate growth explains industrial expansion and diversification in terms of an accommodation to a growing and differentiated demand for that industry's goods and services.

Over time, the market demand for any given type of industrial product usually does not grow uniformly. Over and over again it has been observed to follow a certain dynamic, which is called the product's life cycle. If an innovation brings a new product type or quality onto the market, the market accepts the product only hesitatingly at first. After a certain introduction period, the market rushes to acquire the novelty, first with growing and then later with diminishing appetite. Finally, with time the demand for this entity dwindles to a minimum. The cycle is complete.

Figure 2–2 illustrates such life cycles for (first-, second-, and third-generation) computers in Great Britain. In 1959, only first-generation computers were being installed; two years later almost half of the computers were second generation. By 1963 these new models had taken over almost the entire British market. In 1965, however, the third generation of computers appeared; within a half a year they managed to break into the demand for the older models. In 1966–1967 already more than half of the computers installed in Great Britain were this new type; later, there was virtually no more interest in the older ones.

The individual life cycle of a product reflects the sales figures over several years. This cycle is illustrated by a curve that is relatively bell-

Figure 2–2. Life Cycles for First-, Second-, and Third-Generation Computers *(market shares in Great Britain: 1959–1967).*

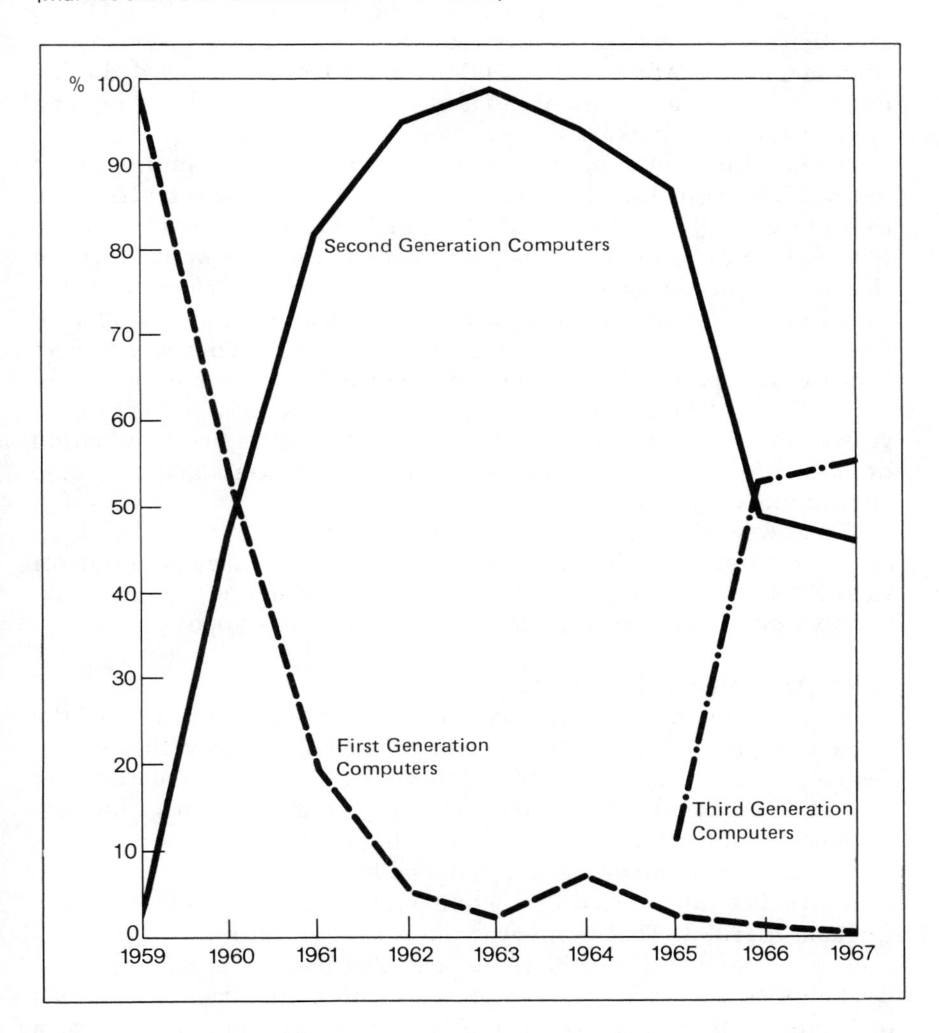

shaped. The life cycles of products introduced in various years overlap. This overall effect has two aspects. If sales fall off, the company will try to introduce an improvement innovation for this product type; the improved products then enter into competition with the older versions. In this way firms often put themselves at a disadvantage, however, without being able to avoid it. The long initial period that a new product requires to penetrate the market forces firms to introduce their new products as early as possibly. Moreover, fear that a competitor may preempt the new market intensifies the push to introduce the new product early. The consumer experiences this process particularly strikingly in the automobile industry where at certain times the consumer is offered both old and new types of cars simultaneously. When only pseudo-innovations are involved the life cycle of a product shortens, that is, the length of time it remains in style amounts to exactly one season.

Product life cycles are generally set into motion by improvement innovations. Usually, life cycle following life cycle will be triggered by the seemingly continuous flow of improvement innovations, and *it is the series of improvement innovations that give momentum to the growth process in the industrial branches.* What Schäfer calls the "breathing growth" in the prospering branches stretches itself over one, two, or three decades until the breathing becomes shallower because the series of improvement innovations are mixed with and replaced by pseudo-innovations. One can thus derive a picture of the growth process of a particular line of business from an examination of both the expansion of capacity by fits and starts and the more continuously growing total sales figures.

We now would like to aggregate the life cycles further. The total sales over time for all the businesses in a branch provide input for the diffusion process,[19] which describes how an innovation gradually is dispersed throughout the population broadens its appeal.[20]

The Diffusion of Innovations

We will now make a summary examination of the dynamics of the successive product life cycle. We will not be concerned with the sales figures for individual companies, but we will be looking at the sales of certain groups of companies that supply a market with a modern durable good, for example, television sets.

The starting point for such a synopsis is a basic innovation. Television was first introduced on a commercial basis ("for business") in England in 1936. That event was the basic innovation for this technology. Although in 1936 there had also been experimental telecasts in Germany, the United States, and other countries, widespread diffusion of television among the West European population was a postwar phenomenon, while in the United States the innovation took off already during the war years.

An innovation such as television is usually introduced to consumers by several firms (licensees) simultaneously. These companies work together to bring the new product into as many households as possible under various brand names. In 1955, only very few German households owned a television set. In 1960, still only every fifth family did. But by 1965 at least 500 of every 1,000 families had one, and today only one out of twenty households does not own a television set. Figure 2–3 shows the dispersion of the television in West Germany during the postwar period.

Holger Bonus, who gathered this data in the 1960s and calculated some prognoses for further dispersion, showed that further expansion depends upon whether market saturation will occur years from now or is already near at hand. This distinction is significant for the following discussion of stagnation as a result of pseudo-innovations.

Figure 2–3. The Diffusion of Television in West Germany from 1955 to 1965 with Different Prognoses for the Following Years.

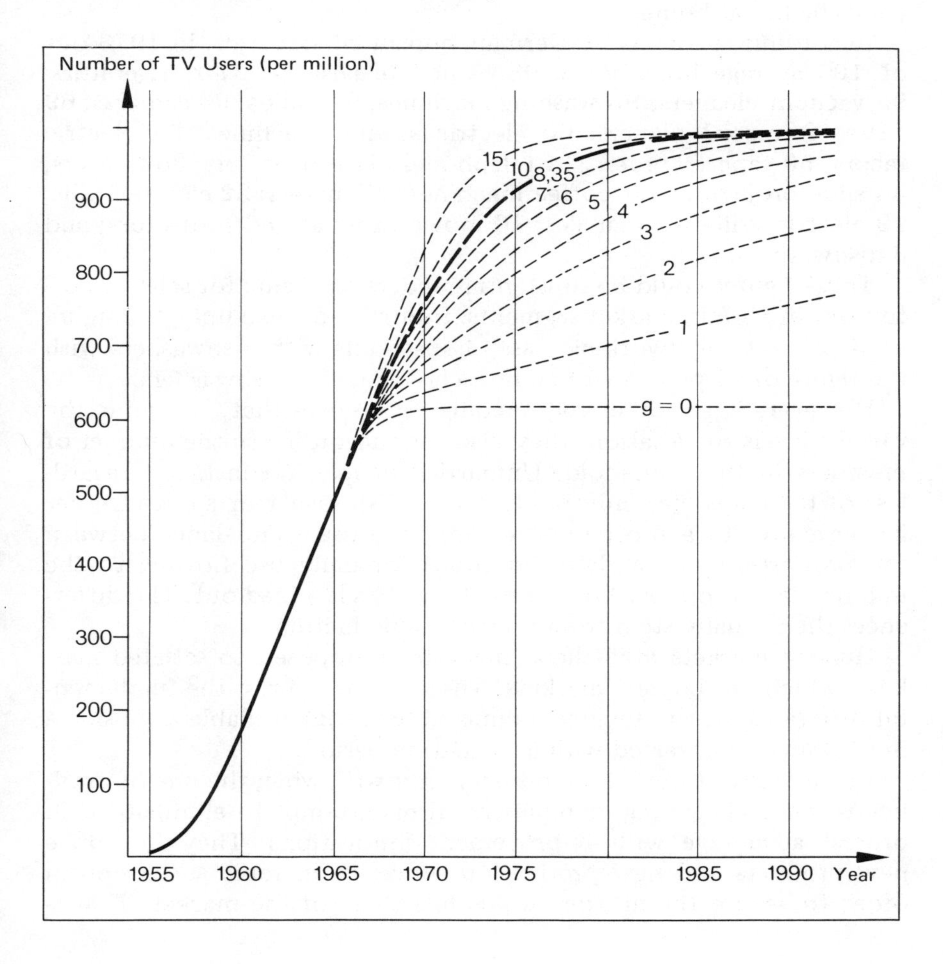

Quality Competition, Pseudo-Innovations, and the Fundamental Cause of Stagflation

Depending upon whether the level of demand on the market is still high or whether it is largely satiated because either very few or almost all of the households with buying power possess the product in question, the manufacturers will plan for quality competition differently. The following figures describe the household possessions of the average German family (wage earner with wife and two children earning a gross monthly income of approximately 1,900 marks). They show the appliances that "Fritz Average" usually possesses. Thus, they show what he is likely not to buy except if the newest products offered in the stores are highly superior to the durable goods he has at home.

According to the West German bureau of statistics, in 1973 out of 100 average households, 98 owned televisions; 97 refrigerators; 95 vacuum cleaners; 95 washing machines; 94 radios; 93 cameras; 69 cars; 62 record players; 60 electric sewing machines; 54 electric razors; 50 tape recorders; 37 telephones; 37 typewriters; 36 freezers; 33 slide projectors; 27 coffee machines; 27 mixers; 22 electric drills; 19 electric grills; 11 mangles; 10 movie cameras and projectors; and 3 dishwashers.

These figures could be substantial understatements for sales opportunities in certain market segments. Contrast, for example, the figure of 3 percent for "working class" households with dishwashers with the figure of 27 percent for all households having dishwashers.

However, it is not merely income differences that determine the various levels of satiation; they also vary according to the number of members in the household. Unmarried people, for instance, regardless of their income, only rarely own a dishwasher or a deep freezer in Germany. In a prosperous society, the usual imbalance between the proportion of standardized goods for daily use flowing to the middle class as opposed to the working class is wiped out. The differences then usually stem from noneconomic factors.

Hungry markets are sellers' markets as opposed to satiated markets, which are buyers' markets. These terms express the fundamental difference in the socioeconomic elbow room available to business in a boom as contrasted with a stagnation period.

In the early stages of an industry's growth, when the line of product is relatively young, competitive firms attempt to eliminate each others' advantage with improvement innovations. They introduce new products and new production processes in rapid succession in order to secure the maximum possible share of the market. That is the Golden Marketing Rule. It promises the greatest market power and the highest return on investment. Process innovations should lower costs and reduce waste so that prices can be lowered if necessary for winning out in price competition. Product innovations should make the quality of the wares so attractive that more buyers will reach for this brand than for the competitors' brand. In this early phase, research efforts in the youthful industry are extremely useful, and new ideas are highly valued because they help to capture a larger part of the market.

In the late stages of an industry's growth, however, valuable ideas for improvements in a product are extremely rare because the usable fund of knowledge has almost all been applied. The obsolescence rate of engineering know-how depends upon the industry's stage of growth. Furthermore, wherever suggestions for improvements are made, they are received more or less halfheartedly by management. This is because in so-called mature markets most products have been standardized into mass production, the equipment has been tailored to this norm, the assembly line set up accordingly, the factory buildings constructed along these lines, and the market strategies clearly planned out. How can one change, rebuild, and rethink all of this? Because the familiar technology has harvested most of the possible profits and the overgrown organizations have become inflexible, the large companies tend to resort to what I call pseudo-innovations. Improvement innovations become increasingly rare as industries reach the limits of that industry's growth potential.

In the early phase of the growth cycle, youthful industries are eager for workable improvement innovations. They are not constrained by their own overblown administrative apparatus, and the size of their operation does not stand in their way. This is not true for the large companies. In the advanced stages of growth, the name of the game is concentration and long-term market control. Large firms then are no longer in the position to swiftly change their products if buyers shift preferences. This is the dinosaur effect.

One of the pioneers involved in intensive research for the past twelve years to evaluate the success of or returns from innovations is my former institute colleague, Frederic M. Scherer. The wealth of experience that he and his co-workers have acquired in investigating the marginal returns from investment in innovations in most of the industrial branches he summarizes as follows: "The increase or decrease in earnings growth from innovations appears to vary systematically with the size of the company. The pattern seen in most industrial firms begins with a phase of increasing earnings growth; in the large firms, this gives way to a phase in which the growth of earnings slows down; and there have been certain hazy indications that the largest giants will then again experience increased growth."[21]

There are several concepts buried in Dr. Scherer's careful formulations that should be kept distinct. On the one hand, a *growing* company is typically the *most innovative* form of enterprise as long as the demand for a type of good is still growing and as long as shares of the market can still be captured by quality competition. Potential buyers are still reacting to quality improvements and to reduction of prices through efficiency innovations. Thus, the companies that have not reached their full growth will continue to introduce a variety of improvement innovations. On the other hand, the *large* corporations (with the exception of certain industrial giants) are *less innovative*, relatively speaking. If the market has reached its growth limit and the major market segments are in the firm grasp of a few large corporations, companies will show little interest in engaging in strong quality competition. The satiated market will not react elastically to improvement innovations, nor is it worthwhile for smaller competitors to do battle against the immovable bastions of the large corporations. There are relatively few exceptions to this. After Michelin brought out the steel-belted radial tire in 1971, its market share jumped from 14 percent in 1971 to 25 percent at the end of 1974.

A positive correlation between company size and innovativeness occurs only in high technology where you have no small firms with which to compare innovativeness. These are the regulated areas such as defense, mail and communications, and atomic energy where the government is the major purchaser, and where only a few companies can offer the goods in question. The amount of tax money given to these giant firms as state subsidies for research, development, production, marketing, insurance, and other follow-up costs clearly is a decisive factor in what seems a high rate of innovation by the giant corporations.

It is important to note that in advanced stages of branch growth, improvement innovations are replaced with increasing frequency by pseudo-innovations. A number of factors account for this slowdown of technical advance. Partly this is because the fully grown large companies cannot move ahead rapidly in quality improvements since no prospects exist for worthwhile improvement innovations. Partly the large firms in nature technologies simply no longer want to engage in innovative competition with the same intensity as before. And as market power, exclusive rights, governmental regulations, and other mechanisms erect barriers to entry into the market, the need for offering highest quality or best service diminishes.

In markets for technology-intensive, standardized goods for daily use, the reality is that outsiders will have little chance to participate in the game of dividing the market into market segments that then become exclusive spheres of interest. Usually, no small capital investor can afford the X-million dollar admission ticket required for a market entrée. And what major investor would be so foolish as to pour money into a product line where a large percentage of the facilities are already not being employed to capacity? The large corporations that settle into such mass production markets are formally an open club, but in reality no one else can gain admission. In Britain, Germany, Japan, and in the United States, many markets have become highly concentrated. It is against this background of virtually closed markets in the types of goods that we associate with a prosperous society that we must examine the problem of restraints on competition. We will not deal with the barriers to market entry any further. However, we will now examine two other effects of market power in some detail, the effects on prices and the effects on quality. We will begin with the latter.

In recent years, many branches of concentrated industry have reached their growth limits. In order to gain a sense of the volume of pseudo-innovative activity in these industries, we (R. Heitmeier and I) focused on the area of consumer goods, examining the 120 groups of products that the Warentest Foundation in Berlin had tested during the eight years prior to 1975. In order to perceive quality changes in these products, we had to limit ourselves to the sixty-six product groups that had been tested two or three times. The test results are documented in *Test* magazine, which in Germany is read by 4 million consumers.

Comparing the results of any two successive tests, three quality characteristics were observed:

1. Functionality of the product
2. Safety for the user
3. Durability of the product

The results showed that most of the qualitative improvements, that is, in twenty-five of the sixty-six product groups, resulted in increasing the products' safety. Functionality was improved and durability increased in twenty-three product groups. (Durability fell in only two groups, a fact that raises questions about alleged planned obsolescence.) The circumstance of particular interest for our purposes is the lack of any change in the functionality in forty-three product groups (65.2 percent) during recent years.

Durability and safety are qualities that the user discovers only after the product is purchased whereas functionality, that is, what the product does for you, can be gauged before purchase. Thus, functional improvements are directed toward potential buyers as well as the competitor's regular customers. Since we found no improvements in functionality in two-thirds of the product groups, we can infer that quality competition in the majority of consumer goods markets is at a very modest level.

It is not easy to determine whether quality competition becomes so attenuated because it cannot be more intense when companies are reaching their technical perfection (the depletion effect) or because it has been deliberately choked off. The twenty-three (out of sixty-six) product groups with functional improvements provide some indication as to which of the two hypotheses explains the facts. The functional improvements that were observed fell into two groups. The follow-up test established a qualitative improvement in fourteen product groups that had received particularly poor quality ratings in the initial test, which *Test* magazine had then widely publicized. (Examples include bathroom scales and electric irons.) As a result, the manufacturers raised the quality level dramatically. In nine product groups the quality rating rose from good to better; the products involved were mainly products that relatively few households owned as yet. (Examples included automatic slide projectors, electric grills, and dishwashers.) For these goods there are still market shares to capture, and improvement innovations are one means to that end.

The producer whose market strength or specialization allows the producer complete dominance in a market segment uses pseudo-innovations as a means to maximize profits. In the segmented market domains of an oligopoly, the buyer is left only with a small choice of product categories; within the categories the buyer is constrained by the prices set by the producers. Given the separation between market segments and the way in which each is provided with pseudo-innovations, it is possible for the producers to keep prices comfortably increasing without resorting to illegal price fixing or cartel arrangements.

This mechanism of replacing the improvement innovations with pseudo-innovations secures the bastions of the established producers in many industries and makes the elimination of price competition an accomplished fact, which in turn makes illegal or shady pricing practices unnecessary. The antitrust lawyers and governmental cartel watchers have not yet taken this situation into full consideration when they seek out illegal and unfair price fixing practices. This is why so many antitrust suits against major corporations recall Don Quixote's battles against the windmills.

Therefore it is not surprising that inflation and stagnation appeared simultaneously in the industrialized nations because many of the modern growth industries in these countries were beginning to sense limits to their growth at approximately the same time, namely, after the mid-1960s. The common cause for inflation and stagnation is an open question in modern economic theory. Keynesian theory cannot provide an explanation for stagflation because this macroeconomic theory does not differentiate between exceptional developments in individual sectors.

Stagflation is a Syndrome of Prosperity. When fully grown industries begin to notice the limits to their market's growth because the consumers are nearly fully supplied and there are fewer and bigger competitors remaining in the market, for the large companies conditions then become ripe for restraints on quality competition. Innovation is expensive and upsetting in a segmented market, and a rise in price can be justified with a mere semblance of innovation. Although during the expansionary phase there was pressure exerted by improvement innovations to bring costs and prices down, and quality up, there is no similar pressure in later phases. Now, pseudo-innovations can be and are being employed in many market segments as an unimpeachable tool for justifying price increases, supported, of course, by appropriate sales promotion.

One can see why the conventional theory of inflation is inadequate when one notes the lack of any comprehensive economic theory of advertising. National economists with so-called liberal leanings see advertising as a necessary evil while they complain about the waste of scarce resources or the sly seduction of clients. They do not seem to realize that this apparent waste and the need for stimulating demand are conditions requiring simultaneous explanation. Economists with so-called social leanings see advertising, once it exceeds its informative function, as an instrument for the exploitation of the consumer, which with the help of aesthetic innovations brings more profit to capital.[22] This latter view exaggerates the role of pseudo-innovations whereas the former view underrates it.

The inaccurate appraisal of the pseudo-innovation—its undervaluation by liberal economists and its overvaluation by social economists—provides the key to the main cause of stagflation. In the stagnation phase of the growth process of an industry the inelasticity of demand allows the supplier to substitute pseudo-innovations for improvement innovations, and thus the buyer does not see the lack of real quality improvements in the products and services for which the buyer is asked to pay higher prices (inflation). This is how stagflation begins in the industries with inelastic demand and noncompetitive supply.

One phenomenon that reinforces both the inflationary development and stagnation takes effect when the ultimate limits to growth in mature industries constrain the whirlwind expansion that had previously given momentum to the entire economy via the multiplier effect. Once unleashed, stagflation reveals itself as a self-escalating process affecting significantly more economic branches. One branch may be directly infected by the stagnation in the leading industries. For example, the 125,000 suppliers of Detroit's automobile industry will suffer when automobile production stagnates, and they can only stay in business if they require higher prices for the lower volume sold! As cars become more expensive, fewer cars will be bought. Other branches in the economy will be infected by the profit squeeze caused by the rising prices of capital goods of raw materials, and by increasing wages (the cost-price spiral). Glyn and Sutcliffe almost gleefully describe the profit squeeze in Britain.[23]

Both stagflation mechanisms are intertwined. Chapter 3 will discuss how this process heats up so that the worsening stagnation produces an intensified stampede in prices and vice versa.

The Law of Diminishing Returns Governs Improvement Innovations

Having taken a brief glimpse at our theory of stagflation, let us now return to the problem of understanding the structural change of the economic system. If one examines quality competition in a particular area from a long-range perspective, the following fact emerges. The intensity of quality competition gradually diminishes as innovative development extends beyond the basic innovation into a series of increasingly meaningless improvement innovations. The large companies no longer spur each other on with the fear of a surprise quality breakthrough. In areas of high perfection, there are fewer innovative ideas left to be discovered, and only few of them can justify readjustment of the mammoth production operations. On the other side, very few of the product improvements that are actually implemented are able to generate more demand for the product. This is a frustrating time for the technicians and engineers in the overexpanded technologies and a time when the prosperous citizen begins to abhor the idea of further growth.

The law of diminishing returns governs the series of improvement innovations that emerge theoretically and in practice in a technological area set up years ago by a basic innovation. The law of diminishing returns states that the next improvement will have a smaller beneficial effect than the previous one did. This depreciation of successive improvement innovations continues to the point where the innovations become mere pseudo-innovations. This phenomenon has important repercussions for the entrepreneurs and investors who have committed their managerial talents and capital to an area during its growth phase. At some point, they will no longer want to continue investing in the implementation of improvement innovations that are progressively more ineffectual. This point marks the beginning of the wave of capital migration. This is also a frustrating period for financial and sales managers in these overexpanded companies as well as for the owners of growth stocks.

The key factor creating this situation is the unsatisfactory return on investment. The profits that entrepreneurs and stockholders could reap from improvements or even expansion of the existing operation cannot compare with the profits that the financial market can offer. The liquid capital derived from the sale of an operating large-scale business will yield more profit through financial transactions on the international money and capital markets than it would if it were reinvested in the industrial branch where it originated. The only limitation here is that most businessmen are very reluctant to play the role of speculator.

The dwindling possibilities for improvement in a mature technology cause the investment opportunities in this industrial sector to also dwindle. Product innovation in the sector comes to a halt if it does not produce completely new sales ideas, and the whole branch then stagnates.

Not long ago, the growth curve for any given industrial branch might seem to have been growing indefinitely; now, however, growth has leveled off. The investment capital that had accumulated in this branch and until now had always paid for the creation of new jobs today has a strong tendency to migrate. But where will it move? That is the pertinent question in business circles today.

Safety-Valve Export

Long before production capacity in a branch becomes so large that the demand for its output is in danger of being quickly satiated, the salesmen begin to look for sales opportunities abroad. This is a typical tactic for a company that, although still fully engaged in the contest for expansion, is beginning to feel limits to its growth in domestic markets. Raymond Vernon discovered that the trend toward exporting exhibited by various growth industries follows a distinct pattern.[24] It is influenced by whatever improvement innovations are still available to temporarily widen the shrinking technological gaps between domestic and foreign products as well as by the material costs and wage differentials evident in the different countries.

With his unique perspective on modern industry, Vernon observed a change during the 1950s and 1960s from exporting to importing in the United States and a related change from importing to exporting in the other industrialized nations such as Great Britain, Japan, and West Germany. He explained this phenomenon by using the life cycle theory and accurately depicting the situation in many maturing industries (Figure 2–4). During that period, the growth race was still rapidly continuing.

In the 1970s, with improvement innovations being replaced by pseudo-innovations more freely on the path to a technological stalemate, the clear-cut pattern in the international division of labor blurs again. There is evidence of the export-import pattern reversing in mature technologies. For example, the United States and Canada increased their world market share of automobile production from 32.6 percent in 1970 to 34.4 percent in 1976, while the share of West European automobile makers dropped from 39.8 percent to 32.1 percent between 1970 and 1976.

During the 1950s, American companies enjoyed a technological advantage. As the first group of industries to develop new products (from A as in automatic drive to X as in xeroxing), they were also the first to feel the need to export. Companies in the other industrialized nations could not make use of this strategy until they reached a similar level of technological advancement in the sixties.

The Americans dealt with foreign sales much differently from the Japanese and Germans when they began to export. United States corporations established foreign subsidiaries and bought foreign companies as if they were souvenirs. With these corporate offspring, American companies built up so-called multinational firms, empires on foreign soil. German and Japanese firms did not venture with the same boldness. Although they established branches abroad, these branches were primarily distribution centers, and production was usually carried on at home. A striking term for the buildup of multinational business concerns, the subtle modern variants on the imperialist strategy (Lenin, Luxemburg), "The American challenge," which was invented by Servan-Schreiber, a Frenchman.[25] It is not surprising that the reversal of this trend—the infiltration of the mar-

Figure 2–4. Phases for Life Cycles of Products.

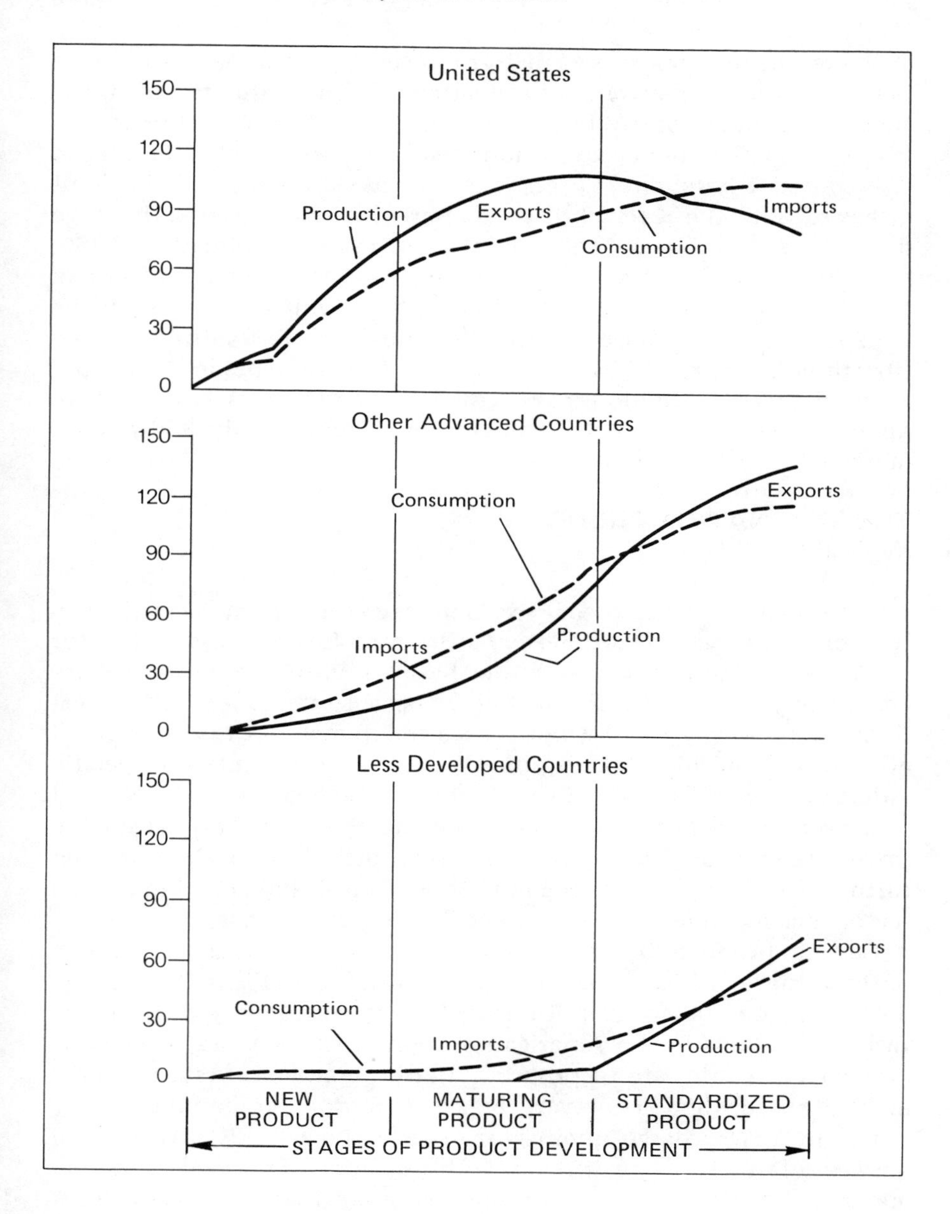

Source: Raymond Vernon, "International Investment and International Trade," *The Quarterly Journal of Economics*, 1966, p. 199.

ket with cheaper products—was quickly labeled, "The Japanese Challenge."

While the export-import pattern reverses, firms have to decide between producing at home or creating foreign production facilities, that is, to build or to buy. This decision, of course, depends on wages. By 1976, one average industrial work hour costs 16 marks in the United States but 17 marks in West Germany, and the joint effect of both the reversal of the export-import trend and the wage differences is visible in direct corporate investments abroad. In 1965, U.S. firms invested 1.5 billion marks in Germany, whereas German firms invested only 69 million marks (then worth 25 cents) in the United States. One decade later this imbalance has shifted. In the first three quarters of 1976, German direct investments in the United States reached 1.3 billion marks (then worth 40 cents), whereas U.S. direct investment in West Germany amounted to only 356 million marks.

THE STAGNATION TREND IN GNP GROWTH

The product life cycle theory explains stagnation in individual industrial branches because of the exhaustion of further possibilities for technological improvements, either because all of the new expertise has already been put into practice or because improvement innovations are no longer worthwhile in a saturated market or both. We will now show how this vanishing of investment opportunities in specific industries can infect the entire economy. Stagnation in what until now were the most flourishing areas of the economy causes the slower tempo of activity in the other branches to slacken even further. Reaching the upper limits to growth in individual industrial sectors has repercussions that affect the economy as a whole.

Certain product life cycles illustrate how sectors can simultaneously achieve a high level of market saturation (Figure 2–5). The data have been taken from Table 2–2; they concern the diffusion of various products in the American economy. Unfortunately, because monetary statistics are of primary interest, commodity statistics are difficult to gather. However, Figure 2–5 does communicate our point. In 1978 a large number of industries have only a very small market left to supply with commodities. In earlier decades, the demand pull from the unsatiated markets intensified the level of economic activity; this activity has now slackened considerably. The resulting stagnation in individual demand areas inevitably penetrates the entire economy.

Figure 2–5. The Diffusion of Various Products in the Economy *(for data see Table 2–1).*

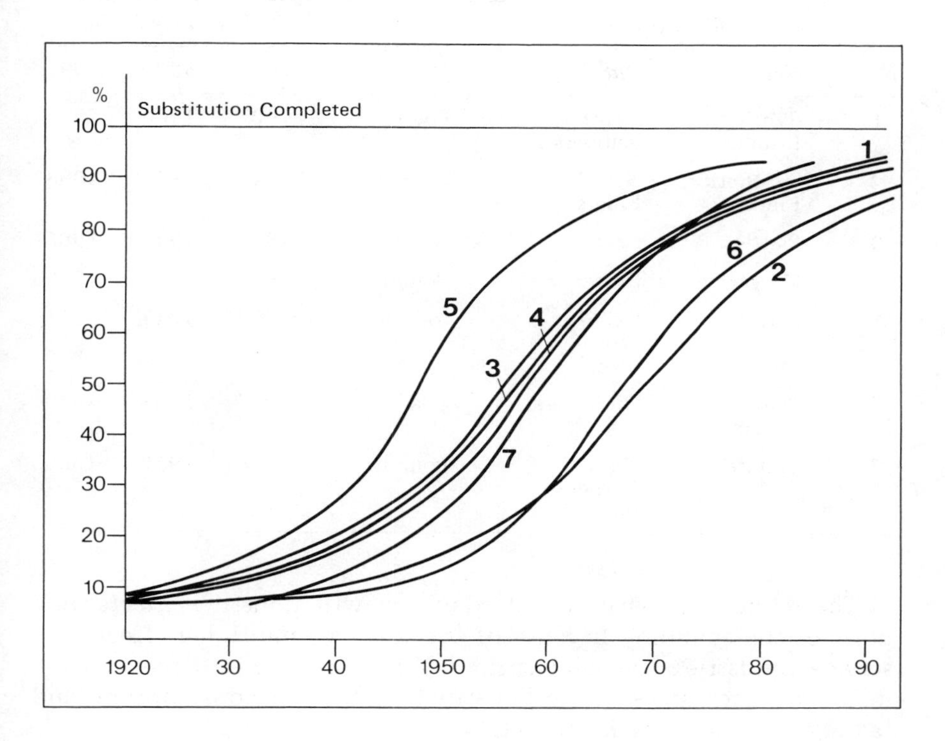

The information in Table 2–2 requires additional explanation if we are to understand the dynamics involved in the distribution of new goods throughout the economy. New goods may also supplant older products. Thanks to Dr. Pry of General Electric we have some statistical information about these substitution processes.

We have very little information about the residual demand on the markets for mass-produced standardized goods such as automobiles, refrigerators, textiles, railroads, airplanes, and so on. However, it is widely known that many of these sectors currently are not using more than three-fourths of their production capacities. This phenomenon is evidenced by the unfortunate fact that shortened hours and permanent layoffs have become normal occurrences in these branches; and that despite these measures the companies are still losing money. Specifically the world capacity of synthetic fibers (second on the list in Figure 2–2) far outpaced the growth of demand for this substitute for natural fibers.

Table 2–2. Processes of Technological Substitution.

No.	Substitution: New	Substitution: Old	Units of Mass	Exchange Status: 10%	Exchange Status: 50%	Exchange Status: 90%
1	Synthetic Rubber	Natural Rubbers	Pounds	1927	1956	1985
2	Synthetic Fibers	Natural Fibers	Pounds	1940	1969	1998
3	Plastics	Leather	Hides	1929	1957	1985
4	Margarine	Butter	Pounds	1929	1957	1985
5	Electric-Arc-Steel	Oven-Stove-Steel	tons	1923	1947	1971
6	Water Colors	Oil Colors	gals.	1945	1967	1989
7	Sulfate	Plant Turpentine	Pounds	1938	1959	1980

The reduced activity in yesterday's growth industries infects wide areas of the economy because of the so-called multiplier effect. The service industries stagnate; shrinking incomes lead to reduced demand in other areas—from construction to beer consumption; and tax revenues decrease in real terms.

This brings us full circle in the developmental cycle, which over hundreds of years stimulated socioeconomic evolution to fashion toward its present state. The cycle begins with a technological stalemate resulting from stagnation in the formerly most highly developed industrial areas. This situation engenders the cultural, political, social, economic, and technological conditions required for the emergence of a cluster of basic innovations. These innovations establish numerous new manufacturing and service industries, which through successive improvement innovations then become increasingly useful for ever larger segments of the population. These branch cycles run in tandem over the long term, exhausting themselves at approximately the same time in a new technological stalemate as the series of improvement innovations governed by the law of diminishing returns produces progressively less significant improvements. These improvements eventually become pseudo-innovations, a process that, as we have shown, manifests itself in the overall economic system as both stagnation and inflation.

We will now attempt to outline a picture of the overall economic trend derived from the multitude of interconnected branch cycles. We are, of course, fully aware that we have neither the procedural tools nor the substantive data necessary to conduct a formal study. Furthermore, we have only a theory of complementary goods or intercoupled purchases to explain why different goods innately belong together to such a degree that the growth cycle in one branch produces market effects that carry over directly to the growth cycles in other branches and ultimately cause these highly elastic cycles to run parallel courses. Until now changes in human life-styles have been considered to be a subject of research for anthropologists rather than for economists. Without this important research, we know almost nothing about the sensitivities involved in the "complementary substitution"[26] of sets of new goods for sets of old products and services for the average consumer.

The empirical gaps in the fund of knowledge available to economists are matched by the methodological weaknesses in an attempt to aggregate microprocesses into one macroprocess. The mathematical difficulties are comparable to those encountered by Einstein in his struggle to integrate partial difference equations into his general description of relativity theory. Einstein comforted himself by saying: "God does not care about our mathematical difficulties. He integrates empirically."[27] We will keep this statement in mind as we attempt to view the process of growth in the economic system as an aggregation of the processes going on in the individual branches.

The Stagnation Trend in West Germany and the United States

We will now demonstrate how long-term economic progress in West Germany and the United States has followed an *S*-shaped growth curve that appears to be continuing in the same configuration. We maintain that the economic development in these countries provides a pattern for other nations. This assertion is partially founded upon our awareness that the *S*-shaped trend must be valid for the American economy if the cyclical argument is to hold any weight. The American economy is the most influential variable in the Western system both because of its size and its leadership position. We will also outline the *S*-shaped trend in West Germany because the postwar development of the German economy is extremely atypical as the 1974 Licari-Gilbert study of growth rates in the OECD countries showed.[28] From the German example, one can then infer that the *S*-shaped trend is a typical phenomenon for all other cases, given that it was found to apply in the least typical case.

At this point in an earlier version of this chapter, we outlined the arguments for working out an aggregation of the branch cycles. One important aspect of this argument was the hypothesis that these branch cycles logistically run in *S*-curves. Such microcycles have linearly declining growth rates that one can add up to see that the aggregate macrocycle must also have this *S*-shaped curve. However, I have since found this aggregation formula verified and thoughtfully worked out in Professor Tintner's writings and shall therefore simply refer to it here.[29]

Our purpose in demonstrating the *S*-shape of the trend in the American and German national products is to illustrate that the current economic troubles fit into the long-term trends in economic development as the Kondratieff cycle depicts them (see Figure 2–8). Figure 2–6 shows the *S*-shaped trends in the American national product and private investment as Professor Tintner estimated them in 1972. Substantial time has elapsed since then. My newer estimation takes account of more recent data (up to 1975) on the subject. This accounts for the differences in the two estimates in Figure 2–6.[a] The more recent (1975) estimate shows the *S*-shaped curve much more clearly, and in the years since 1975, the *S*-shaped trend wall depicts the leveling off of recent growth. The *S*-shaped trends in Figure 2–6 are an impressive illustration of how the branches of private American industry reach a plateau at similar levels and how overall economic production progressively falls off. The growth of the German national product reveals the same phenomena, albeit less markedly (Figure 2–7). According to the Licari-Gilbert study, the growth pattern for the national products in other industrialized nations should resemble the American situation more than the German one.

We have now demonstrated that the limits to growth in individual areas of the economy have an effect on the system as a whole. The whole economy must also be considered to possess limits to growth if it preserves its sectoral structure. When new basic innovations do not occur, overall economic growth gravitates toward zero, sometimes even causing a depression with its accompanying negative growth rate. Only the new opportunities offered by basic innovations can prevent a depression, or if necessary, overcome its effects.

The form of the *S*-shaped trend derives from the diffusion of innovations throughout the leading industries; it therefore runs a very similar course in all of the industrialized nations. How does this trend

[a]Tintner's log-normal approach places more emphasis on the starting points than my logistical estimation approach does. The latter approach focuses more on the end points.

Figure 2–6. *S*-Shaped Trend in the Gross National Product and Private Investment in the United States.

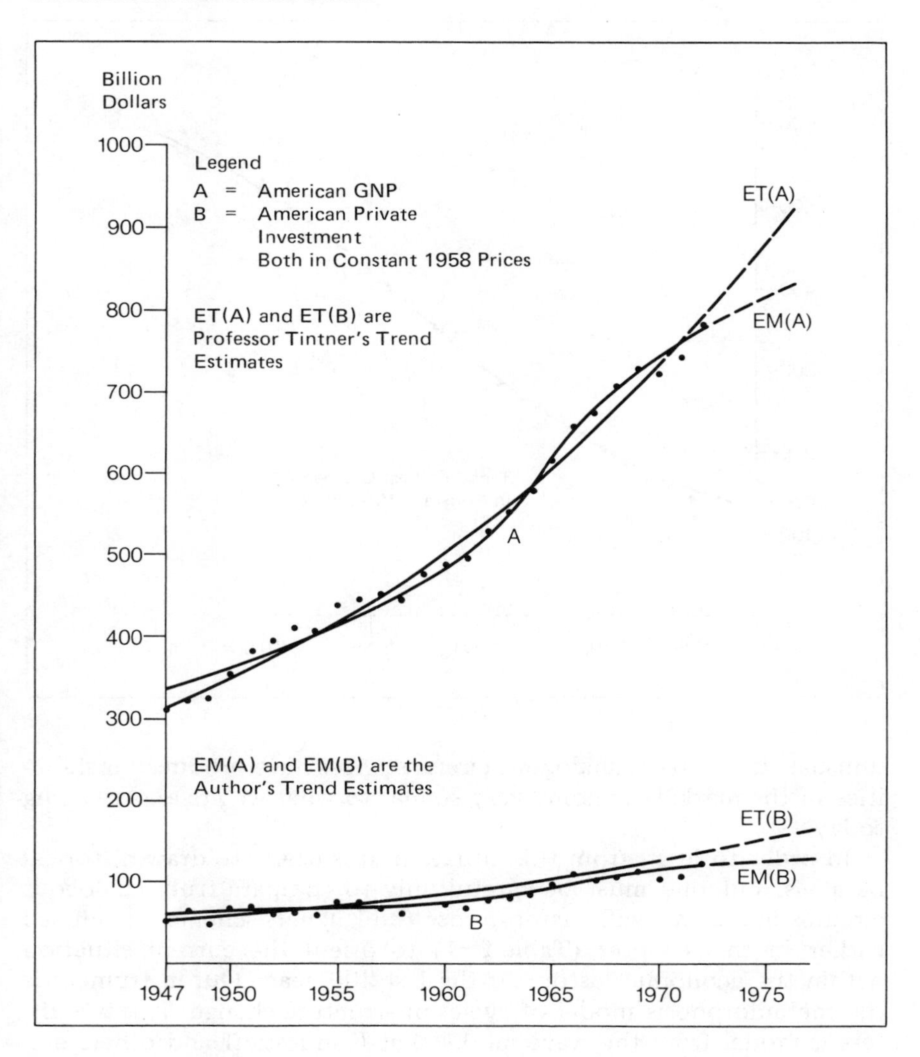

fit into long-term economic development during the 200 years of the industrial age?

The Changing Trend since the Industrial Revolution

We must remember that while today's crisis (the technological stalemate) may be unique in its details, its structured features are not

Figure 2–7. *S*-Shaped Trend in West Germany's National Product.

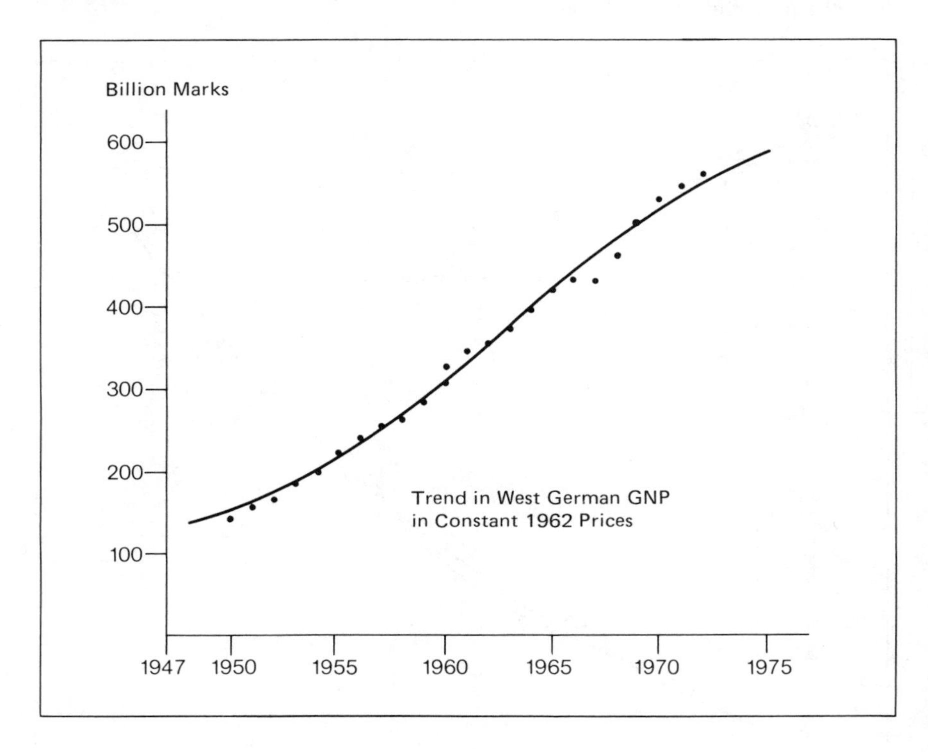

unusual. In the past, analogous circumstances have produced instabilities in the market economy very similar to what we are experiencing today.

In order to learn from this situation, it is useful to draw historical parallels, but one must be careful only to compare truly analogous circumstances. We will therefore use the Kuznets schema introduced earlier in this chapter (Table 2–1) to orient the current situation within the economic history of the last 200 years. Our instrument is the metamorphosis model of cycles of structure change. This is a different model from the wave model that Kondratieff and others employed.

If one, however, agrees with Kondratieff, Kuznets, Schumpeter, and the other economists who see these changing economic trends as a series of long waves, one can use the Kuznets schema to produce a wave configuration like that shown in Figure 2–8. While the *wave model* distinguishes phases (prosperity [*P*], recession [*R*], depression [*D*], recovery [*R*]) that have been reiterated over the past 200

Figure 2–8. The Metamorphosis Model of Industrial Evolution.

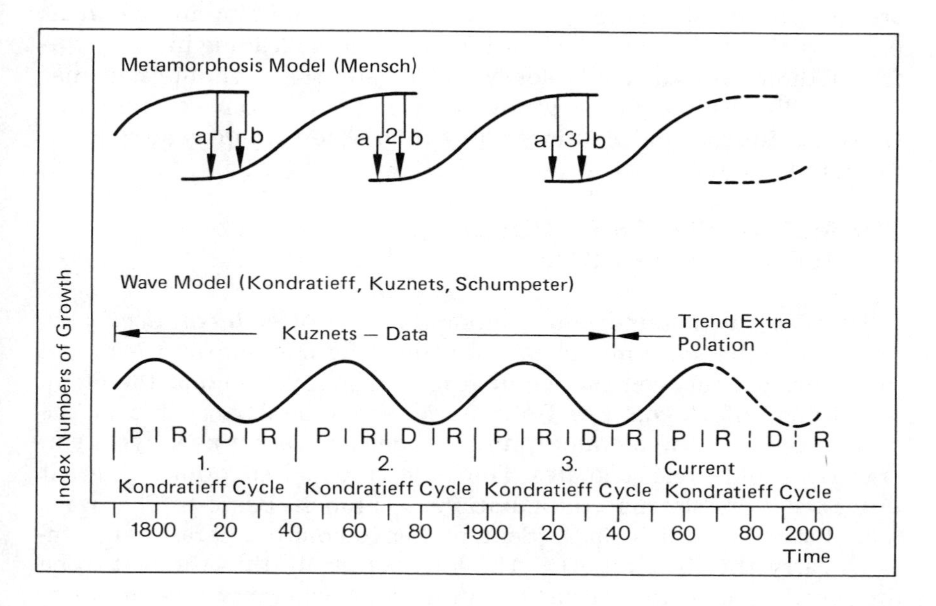

years, it sees them as part of an uninterrupted sine curve with repetition of the P, R, D, R sequence at regular intervals.

We have abandoned the notion that the economy has developed in waves in favor of the theory that it has evolved through a series of intermittent innovative impulses that take the form of successive S-shaped cycles (see the upper portion of Figure 2–8). The cycle that I think we are currently passing through, which is pictured in Figures 2–6 and 2–7 as an S-shaped curve, fits perfectly into this schema, which I call the *metamorphosis model* depicting long periods of growth and relatively short intervals of turbulence.

Socioeconomic development clearly does not proceed in a continuous flow; there are breaks and upheavals over time that cause significant variations in its tempo. Despite this uneveness, the total process exhibits significant regularities; apparently these regularities have more force than policymakers would like us to believe when they contend that a degree of freedom exists for active direction of the economic process. There is a large difference between regularity and historical determinism. The wave model incorporates a deterministic recurrence of phase transition; any metamorphosis model does not. It allows for speedup and slowdown of change.

In times of a sluggish economy the interplay between stagnation and innovation offers opportunities for persons and groups with initiative. At the same time that wide areas of current economic interest are gripped by stagnation, creative progress is building in new areas of activity. The industrial evolution opens doors leading in many different directions. In technological stalemate, the economy becomes structurally ready for basic innovations that could go in several directions. Although some are more likely than others, there is no determinism in this process.

THE METAMORPHOSIS MODEL IN THEORY AND PRACTICE

What does our metamorphosis model have to offer *theoretically* beyond what the wave model can provide? The metamorphosis model differentiates between the latent forces of innovation and the manifest forces of stagnation. Both forces are indistinguishable in the wave model. In the metamorphosis model one sees a new cycle emerge as the old cycle stagnates. Thus, the model gives room to "latent processes" or "latent configurations" (Francis Bacon). The wave model stimulated Joseph A. Schumpeter to coin the phrase "innovations carry the Kondratieff." The long waves are thus thought to be produced by a positive thrust of innovations, whereby these impulses are thought to always outweigh the negative stagnation phenomena in times of above zero growth. Stagnation is thus canceled out of the wave equation.

The metamorphosis model incorporates stagnation explicitly as a separate force. This model can therefore encompass both the effects of stagnation on innovation and the feedback with which innovation influences stagnation. Within the socioeconomic context, the entire evolutionary process is bound into a self-regulating cycle. Stagnation in certain parts of the system as well as in the system as a whole encourages individual improvements in structurally appropriate areas. Innovation, in turn, causes certain formerly valued economic activities to be no longer viable. Innovation and stagnation in part produce one another. The metamorphosis model allows stagnant industrial branches to be separated from new branches that basic innovations have brought into being.

What does the metamorphosis model achieve in *practice*? It offers a calendar of 200 years that is useful for making comparisons between analogous circumstances in different periods.

For centuries it has been said that history repeats itself and that one should learn from the past in order to create a better future. "The use of history is to give value to the present hour and its duty" (Emerson). When students hear this, they might wonder why their lectures on social and economic history offer so few worthwhile lessons. The reason lies in the confusion that change brings with it. Which situation in the past is analogous to today's? When individual details are so dissimilar, how can the general situation be similar? The metamorphosis model can at least partially answer these questions.

The model permits one to infer what the economic climate has been in particular historical periods according to the relative predominance of either innovation and growth or stagnation and crisis. Naturally, there are individual peculiarities in each era that will not reappear. Nevertheless, some structural characteristics and general features of the economic process show amazing similarities, compare, for example, the prosperous years of the recent past, the golden twenties, and the big boom in the 1860s (see Figure 2–8).

More specifically, compare the upper and lower graphs given in Figure 2–9. They depict the direction of structural change in the U.S. manufacturing industry since 1889. Obviously, the years 1918 and 1968 separate two trend phases, respectively, the first ones exhibiting an expansionary type of structural change, the second ones depicting a recessive type of structural transition. In terms of capital investment and labor employment in industry, the analogous phases in the two periods 1900–1918 and 1950–1968 were both capital and labor augmenting, whereas the two periods after 1918 and after 1968 were characterized by labor saving via capital augmentation. The turbulence created by the forces of substitution is what the industrial world experiences during the early phases of a stalemate in technology.

It seems that crises proceed in two stages. At first, certain countries suffered severe business collapses, but the situation did not (yet) degenerate into a world economic crisis. Generally, it seems that crises in the past began in countries that had just ended a war and were not able to cope with the transition from a wartime to a peacetime economy in the prevailing climate of stagnation.

For example, the countries that defeated Napoleon at Waterloo experienced a serious crisis in 1815; because of the particular symptoms that the crisis exhibited, the English labeled it the reconversion crisis. Similarly, half a century later, the economic boom gave way to crisis in 1866, first in France (with the Crédit-Mobilier crisis), next in the United States at the end of the Civil War, and finally in Prussia and Austria after their war. The same story was repeated in 1920 after World War I (see Table 2–4). I have labeled these first-stage crises in 1815, 1866, and 1920 "signal crises." In the early 1970s, when the world economy was again stagnating, the United States' potential difficulty in readjusting from the Vietnam War without

Figure 2–9. The Metamorphosis Model of Industrial Evolution Applied to the Capital-Labor-Allocation in U.S. Manufacturing.

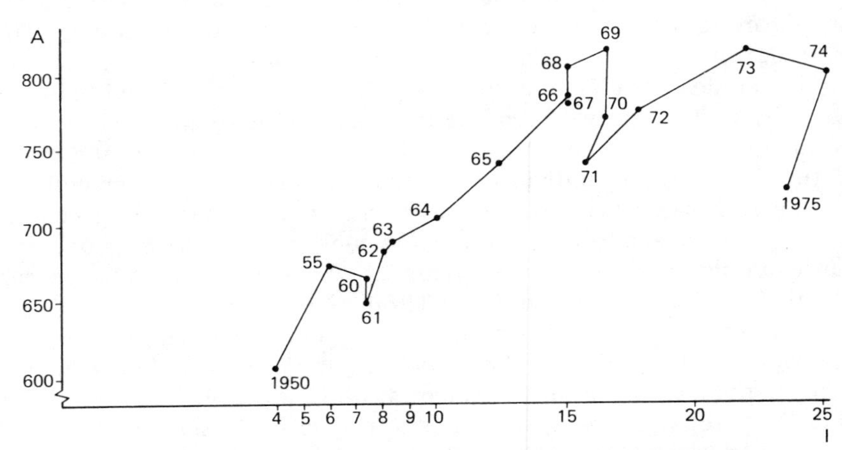

Legend: A = labor input indicator [Bill. weekly manhours]
I = capital input [Bill. of current dollars]

Data source: Stat. Abstracts of the USA, 1976, Series 13o3
ILO-yearbooks, 58er, 66er, 76er.

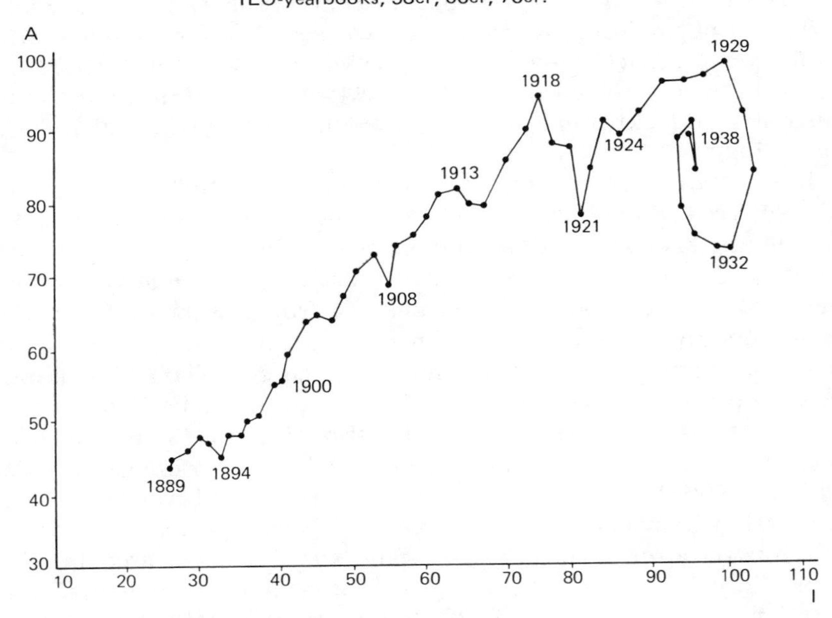

Legend: A = labor input index (1929 = 100)
I = capital input index

Data source: Kendrick, J.W., *Productivity Trends in the United States*, Princeton University Press, 1961, p. 328.

Figure 2–10. Leading Sectors in German Industry 1950–1976.

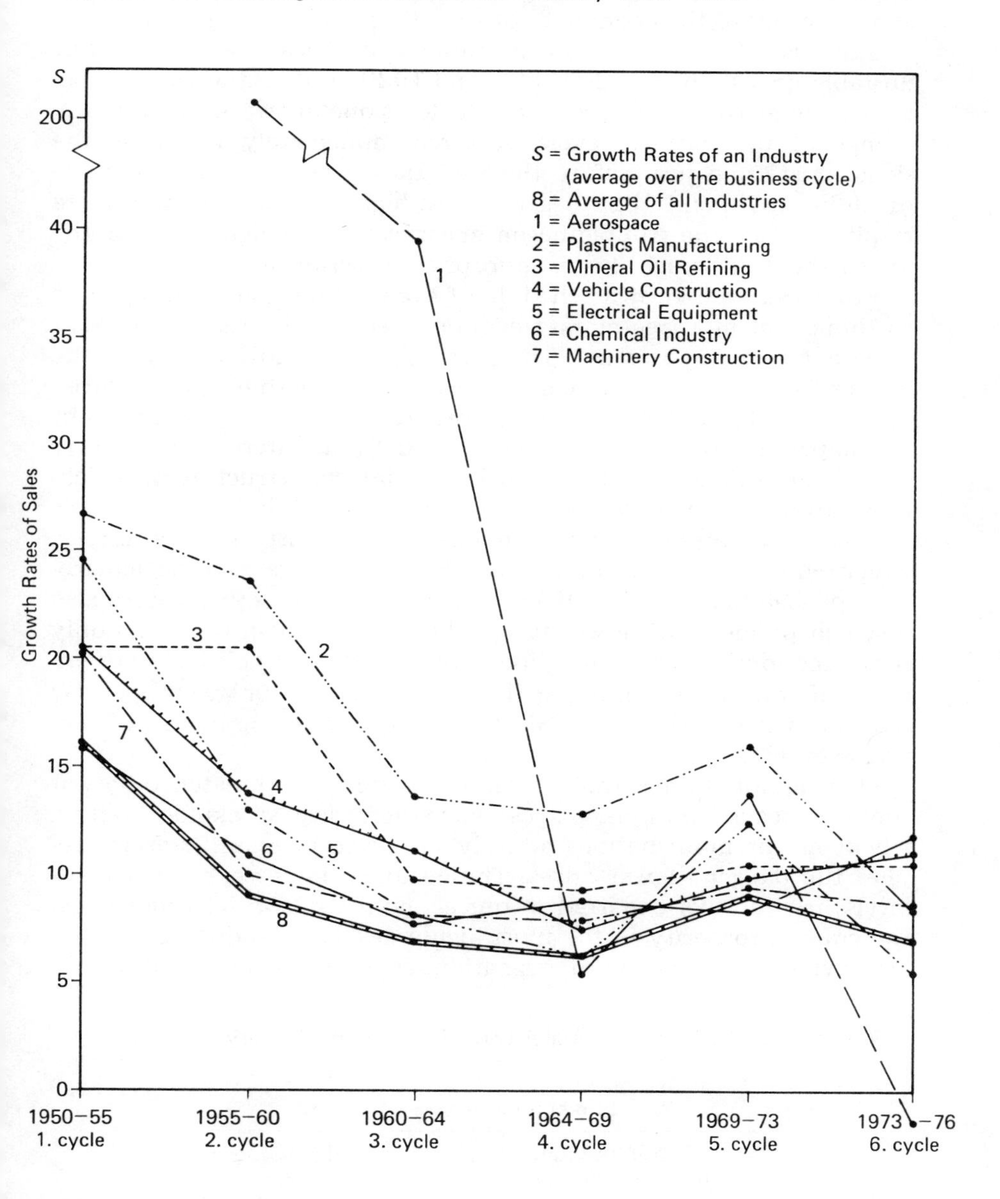

A. Kleinknecht assigned the basic innovations from Table 4-4 to the appropriate branches of industry and found that these branches grew significantly faster than the average of all German industries. However, the growth effect of basic innovations diminishes over time.

a drastic economic slowdown posed a serious threat both to its own economy and to the economies of the other industrial nations.

The crisis timetable set out in Table 2–3 indicates an interlude of unstable growth after 1815, 1866, and 1920 (as does the distance between the arrows in Figure 2–8). At least one or two short-term upswings of the business cycle occurred immediately following the signal crises and just before the well-known world economic crises of 1825, 1873, and 1929 began. These high periods were thus mere respites before the trouble began in earnest. "The higher the top the longer the drop" might be an appropriate description.

From this well-known schedule of events, some people have come to think that in a free market economy, economic crisis is the inevitable consequence of peace. This is an unpleasant notion, and yet the dangers of war are clearly closely related to sluggish economic development. In his peace research report, Klaus I. Gantzel pointed out the appalling correlation and concluded "that international power politics are primarily determined by the internal structure of whichever nations are farthest developed at the time."[30]

Our metamorphosis model delineates the historical periods in which—due to stagnation in the economy—there is a particularly high probability of war. Of course, war can and actually does also occur in periods with lower probabilities. Our model considers only some economic factors and does not include the political determinants of war. Thus World War II does not fit into our war propensity schema, which derives its conjectures solely from the course of economic events.

One cannot surmise that a period is particularly threatened by war simply because nations possess large defensive stockpiles; rather, when one or more nations actively begins to increase its arms supplies the danger of war arises. These sudden increases in armament levels have always occurred in our economic history at times when the era of prosperity in the most highly developed industrial nations was giving way to an era of stagnation. Defense contracts had to sub-

Table 2–3. Time Table of Major Crises in Economic History.

1.	a.	Reconversion Crisis	1815
	b.	World Economic Crisis	1825
2.	a.	Crédit-Mobilier Crisis	1866
	b.	World Economic Crisis	1873
3.	a.	Postwar Crisis	1920
	b.	World Economic Crisis	1929

Source: M. Flamant and I. Singer-Kével, *Modern Economic Crises* (London, 1970).

stitute for sluggish private demands. According to the Kuznets model and Figure 2–8, these inflection points in the past growth trends occurred in 1801, 1858, and 1912. These dates provide us with a guess for the hindsight prediction of which years were particularly vulnerable to an outbreak of war, since the economy-boosting defense contracts that a state in a critical situation would suddenly promote would need to be justified domestically by the engagement in war-threatening behavior abroad. These tactics would soon create a *casus belli* whether or not it was desired.

The timetable of major wars (Table 2–4) fits into the stagnation scheme too well not to cause us some alarm. The economic preconditions for war include defense contracts used as therapy for a stagnating economy. This pattern was repeated before every large-scale belligerent encounter until World War I. World War II does not fit into the economic framework, but the contention that the United States became involved in Vietnam for primarily economic reasons has been discussed too thoroughly to require more elaboration. One can also use this progression with substantial justification to maintain that the years after 1967 also posed a high degree of danger that war would break out. The tragic protraction of the Vietnam War and the eagerness of the Great Powers to supply the Middle East combatants with arms also fit into the schema, as does the historical parade of great "architects of peace"—Metternich, Bismarck, Wilson, and pro tanto, Henry Kissinger. Analogous situations produce analogous men. One can also argue, however, that World War II and the balance of terror have made escalation toward large-scale war obsolete because nuclear weapons have made war unthinkable. That does not exclude the possibility that President Carter asks Congress to approve a 4 billion arms delivery contract in 1978—and gets it in the economic climate, that, according to the metamorphosis model, leaves Congress with little choice.

History teaches that a certain type of economic crisis follows war. After the armistice, people are still suffering from the horrors of war and governments will no longer be able to counteract the prevailing stagnation in private industry with defense spending. Therefore, the economic problem of stagnation will reappear in the postwar period, only in a more aggravated form. The so-called peace economy that was already stagnating before the war must now supply the returning soldiers with jobs. But why should these stagnating industries be more capable of creating jobs in the postwar poverty than they were in the prewar prosperity?

There is therefore an unhappy correlation between the years in which truces or armistices were signed and years in which signal crises occurred: The crisis years are 1815; 1866; 1920 (Table 2–2).

Table 2–4. Stagnation and the Danger of War; Peace and the Danger of Economic Crisis.

Inflection Points in Growth Trend *Increased Risk of War*	*When Major Wars Began*	*Wars*	*End of War*	*When the Signal Crises Occurred*
1801	1799	1799–1802 Wars of the Second Coalition against France; Napoleonic Wars; Napoleonic Wars;	Waterloo 1815	Reconversion Crisis 1815
1858	1854	1854–56 Krimean War (England, France, Italy, Russia); Sardine War 1859; USA 1861-1865; European War (Prussia, Austria, Denmark, German Union) 1864-1866	Königgrätz 1866	Crédit-Mobilier-Crisis 1866
1912	1914	First World War 1914–1918	Compiègne November 1918	Postwar Crisis 1920
		Vietnam War		

The end of the war years are 1815; 1866; 1918 (Table 2–3). What these crises signal is the structural instability of a beginning technological stalemate. In the following chapters we will use the metamorphosis model to show the time correlation between stalemate situations, surges of basic innovations, and shifts away from the depression trend toward economic recovery. Because of the recurrence of structural transformation during technological stalemates we can state with some assurance that basic innovations overcome depressions.

As a final test of the metamorphosis model, we focus upon the employment of foreign workers. This is a problem of great concern to many industrialized nations today. It provides still further evidence for the thesis that similar structural conditions recur in the long-term economic process. The enormous influx of foreign workers into West Germany since 1961 does have a parallel in the preceding Kondratieff cycle naturally, during the analogous economic phase. It also has a strong parallel in the fluctuation of immigration of Europeans to the United States in the last 100 years. Using Figure 2–8, we will follow these parallel developments that began in the 1880s and the 1930s respectively. We will focus on German data.

We should mention at the outset that a depression whose severity matched that of the 1929–1939 depression occurred between 1873 and 1886. It is called "The Great Depression" by Rosenberg[31] and many others. Both periods of economic lows then gave way to similar recovery phases and renewed prosperity. In the postwar period we referred to this recovery as the economic miracle of the 1950s, while our grandparents can recall similar times during the *belle époque* at the turn of the century.

From a structural point of view, what occurred before and during these phases of prosperity did not differ greatly. We can observe a large emigration of Germans during troubled times and a substantial influx of foreign workers into Germany both in times of prosperity and during the first years of a recession. The following data illustrate these phenomena:

1881–1890	1,453,000 German emigrants
1891–1900	544,000 German emigrants
1901–1910	341,000 German emigrants
1907	800,000 foreign workers in Germany
1914	1,200,000 foreign workers[32]

This development two generations ago fully conforms to the change from a high rate of German emigration in the postwar years to the swell of immigration since 1961. While the quota of foreigners in Germany is very high today, it is in no way extraordinary when compared to the large numbers of foreign workers present in Germany before the first World War.

SUMMARY

In this chapter, we have portrayed industrial evolution as a diversification of human ways of life. Over time, the deepening division of labor with its increased efficiency has produced a widening variety of goods and services that, as they are being used by people, shape distinctive life-styles. On the other hand, a number of modern man's problems stem from the deep division of labor, because it fragments people's lives and thereby also weakens the sense of community. For structural reasons and as a result of shifting values, certain goods that once were highly favored may suffer a stagnating demand if they are available in plenty.

Stagnation in the economy is a sign of discrepancies between people's needs and the economy's supply. Firms adjust to changed needs via product and service innovations. Innovations cause changes in the division of labor. Basic innovations open up new areas of activity and consumption, and improvement innovations encourage further technical and organizational progress in these areas. Fragmented production and labor specialization, on the one hand, and industrial and individual consumption, on the other hand, are organized, but only partially harmonized, through market trade. Therefore, despite the general efficiency of the market mechanism, structural and functional difficulties will ensue in special circumstances. They can build up to critical levels at times. Generally, the market mechanism channels innovation and investment into the growth cycles of modern industries, which—given the organizational structure of the economy—will not continue their growth *indefinitely* but will eventually exhaust their momentum. Figure 2–10 shows the effect of basic innovation on increase and decrease of growth in branches of industry.

The phase in the long-term economic process in which the highly developed countries develop into prosperous societies is particularly crisis-prone. This stage signifies nothing more than that many of the markets for standardized goods for daily use have been largely satiated, and thus a number of growth industries will shortly feel constraints from the limits on their market's growth. Satiation in the demand for formerly popular goods and the dwindling possibilities for improvement innovations in many of usual technologies limit the economic prospects of many firms (stagnation).

In the late stages of branch growth, the leading companies have in fact segmented the markets where they operate into individual domains and have discouraged any outside participation. These market segments no longer enjoy the invigorating influence of improvement innovations, which could keep costs and prices down and product quality up, because of pressure from the quality competition. These stratigically separated market segments tend to be flooded by pseudo-innovations and seductive advertising campaigns. Once the quality competition has cooled, these concentrated markets are milked for profit through price increases (inflation).

Rising prices and symptoms of stagnation have the same origin; they strengthen one another and infect progressively more branches of the economy in two ways. The stagnation of real demand leads to attempts to offset revenue decline through price increases, and the increase in factor costs (wages and raw materials) forces firms to raise prices and reinforces the stagnation of demand. This is a true vicious circle (self-reinforcing stagflation).

Waning growth in the leading industries induces overall economic activity to slow down to the point where it is merely stagnating; today this occurs in all of the industrialized nations. On the other hand, the technological stalemate with its depressive tendencies also generates structural readiness for basic innovations. To find basic innovation possibilities that tap the unfulfilled needs of the population is the real challenge of the time.

REFERENCES

1. Simon Kuznets, *Economic Change* (New York: 1953), p. 109–10.
2. Genesis, 2:11–12.
3. Genesis, 2:15;
4. Genesis, 3:7.
5. N. Georgescu-Roegen, *The Entropy Law and the Economic Process* (Cambridge: 1971), p. 127.
6. J.A. Schumpeter, *Kapitalismus, Sozialismus und Demokratie* (Bern: 1950), p. 184.
7. J.A. Schumpeter, *Konjunkturzyklen II* (Göttingen: 1961), p. 1065.
8. Jan Jacobs, *The Economy of Cities* (New York: Vintage Books, 1970).
9. See both works of J. Forrester on urban dynamics and industrial dynamics.
10. G. Mensch (with H. Freudenberger), *Von der Provinzstadt zur Industrieregion, Ein Beitrag zur Politökonomie der Sozialinnovation, dargestellt am Innovationsschub der Industriellen Revolution in Raume Brünn* (Göttingen: 1975).
11. M. Eigen," Selforganization of Matter and the Evolution of Biological Macromolecules," *Die Naturwissenschaften* 58 (1971), 465–523, p. 517.

12. G. Mensch, "Zur Dynamik des technischen Fortschritts," *Zeitschrift für Betriebswirtschaft*, 41 (1971), 295–314.

13. G. Mensch, "Basisinnovationen und Verbesserungsinnovationen," *Zeitschrift für Betriebswirtschaft*, 42 (1972).

14. G. Mensch, "Die Rolle des Ingenieurs in der Wirtschaft von morgen," *Zeitschrift für die gesamte Technik*, 114 (1972), 933–1012.

15. J. Nötzold, "Untersuchungen zur Durchsetzung des technischen Fortschritts in der sowjetischen Wirtschaft," *Stiftung Wissenschaft und Politik*, Eggenberg 1972.

16. DIW, Zur Wirtschaftslage in der Sowjetunion, Wochenbericht 38/1974.

17. J.M. Keynes, *The General Theory of Employment Interest and Money* (New York: 1935), p. 314.

18. J.A. Schumpeter, *Konjunkturzyklus II* (Göttingen 1961), p. 937.

19. E.M. Rogers, *Diffusion of Innovation* (New York: 1962).

20. T. Hägerstrand, *Innovation Diffusion as a Spatial Process* (Chicago: 1967).

21. F.M. Scherer, "Research and Development Returns to Scale and the Schumpeterian Hypothesis" (Berlin: 1973).

22. W.F. Haug, *Kritik der Warmästhetik* (Frankfurt: 1973).

23. A. Glyn and B. Sutcliffe, *British Capitalism, Workers and the Profits Squeeze* (Harmondsworth, 1972).

24. R. Vernon, "International Investment and International Trade in the Product Cycle," *Quarterly Journal of Economics*, 80 (1966), pp. 190–207.

25. J.J. Servan-Schreiber, *Le défi américain* (Paris: 1967).

26. R. Turner Norris, *The Theory of Consumer's Demand* (New Haven: Yale University Press, 1952).

27. L. Infeld, *Leben mit Einstein* (Wien-Frankfurt-Zürich: 1969).

28. J.A. Licari and M. Gilbert, "Is there a Postwar Growth Cycle?" *Kyklos* 27 (1974), pp. 511–520.

29. G. Tintner and J.K. Sengupta, *Stochiastic Economics* (New York: 1972).

30. K.J. Gantzel, System und Akteur; Beträge zur vergleichenden Kriegsursachenforschung (Düsseldorf: 1973).

31. H. Rosenberg, *Große Depression und Bismarckzeit* (Berlin: 1967).

32. S. Geiselberger (Hrsg.), *Schwarzbuch: Ausländische Arbeiter* (Frankfurt/Main: 1972).

Chapter 3

Symptoms of Stagnation

"Time devalues the world"—Horace

Worthwhile achievements take time, but not everything that needs time turns out to be worthwhile. Signs of age in established organizations, a decline in the utility of a commonly used technology, erosion of traditional market positions, cutbacks in industrial branches, exhaustion of dependable sources of raw materials, corporate losses, lack of growth, aversion to prosperity, all of which are symptoms of stagnation, only gradually become apparent. They take a long time to appear.

Stagnation does not usually occur overnight despite some indications to the contrary. The stagnation process in any socioeconomic system interacts with the innovative process. Generally, the illumination from innovations that suddenly appears in our familiar world vision makes us aware of the aging process. This is why old preferences suddenly change as if they were bewitched by a new trend.

Sometimes I use the term stagnation in the plural form to point up the evolutionary interaction between stagnations and innovations. We should be able to distinguish the idea of stagnation as a holistic phenomenon that simultaneously affects the whole and many parts of a system from the notion of a weakening in a particular part of a system. The appearance of stagnation in a pluralistic socioeconomic structure, such as a village or a business, a region or a branch of industry, a nation or a society, depends completely upon whether there has been a convergence of many or of only a few symptoms of stagnation.

Minimal stagnation will always exist in a living world; deaths, dissolutions of organizations, and product depreciation are simply givens. We take all these changes in stride when they occur at a normal rate. However, we consider even a normal number of stagnations in individual parts of the system a stagnation of the whole when this disintegrating trend is not counterbalanced by an adequate rate of creation and growth. Stagnation is lack of innovation.

The interaction between the stagnations and innovations in our working and daily lives is not self-equilibrating; it does not produce a stable balance between these opposing forces. First exhaustion, obstruction, and signs of old age overwhelm an industrial society; then a swell of innovations may return the stagnating economy to the forefront of the industrial world. We perceive a technological stalemate brought about by a destructive flood of stagnations as a structural crisis; however, it is also a time for forces to be marshalled in order to produce a new wave of innovations.

Since the mid-1960s, symptoms of stagnation in the Western industrial nations have become increasingly evident; very few basic innovations were implemented in the years between 1953 and 1973. Moreover, the stream of improvement innovations in established areas of the economy has dwindled, producing progressively fewer useful ideas and being infiltrated by pseudo-innovations. The worldwide business crisis in 1966–1967 was the first serious sign of impending stagnation.

In the following section we would like to expand upon some of the observable causes and effects of stagnation, paying particular attention to the phenomenon of stagflation. Actually, in the last few years it has progressed, as Milton Friedman described it in his Nobel reward lecture, "from stagflation to slumpflation."

CURRENT STAGNATIONS IN INDUSTRIAL BRANCHES

Stagnation of the whole is cured by innovations in its parts. But in what stage of stagnation do the so-called modern branches of the economy in the industrialized nations now find themselves? One can only answer this question individually for each branch. In this vein, the top managers of American corporations meeting among themselves in the "Northpole Conferences," 1971–1973, stated without hesitation: "Much of our present industry is old-fashioned, if not simply obsolete. . . ."[1] Their conclusions stated as early as December 12, 1971 are scathing:

> The U.S. automobile industry relies exclusively on the Otto-cycle engine, now almost a century old. Innovation in the auto industry, upon

which perhaps a third of the Gross National Product depends, has been minimal since the 1930s. The last major innovation introduced was the automatic transmission which became commonplace a generation ago. . . .

Construction is still tied to craft techniques and is the only major industry not to take advantage of mass production technology—at least in the U.S. Many European countries, and even the Soviet Union, have introduced mass production of housing and commercial space on a large scale. . . .

The electric power industry, the largest in terms of capital investment, has seen declining efficiency in its use of fuel for the past decade, reversing a long history of technical progress. The underlying reason for the decline seems to be the fact that the industry has now reached the limits of improvement of the traditional boiler-steam-turbine-generator power plant which has been in use since about 1900. Efficiencies of such plants have now reached quite close to the theoretical maximum; nuclear power plants, which merely substitute uranium for coal, are actually less efficient because of lower operating temperatures, and consequently require larger turbines, cooling towers, and other equipment. Processes of higher intrinsic efficiency, such as magnetohydrodynamics, are not being developed rapidly, nor are new energy sources, such as solar energy, becoming available. . . .

The difficulties of the steel industry, of course, are well known. Here, too, the basic steelmaking technology has not changed greatly since the nineteenth century, despite great advances in the size and efficiency of integrated mills. . . .

Papermaking is another backward industry. A recent OECD [Organization for Economic Cooperation and Development] study concluded that "Pulp and paper-making processes have changed little since the beginnings of this industry, and generally cause considerable pollution. . . ."

Much of the basic chemical industry still relies on older technology. The techniques for making nitric acid, sulfuric acid, ammonia, caustic soda, chlorine, nitrate fertilizer, and many other industrial chemicals date from before the First World War, and once again, except for increases in size and efficiency of plants, have not changed greatly since that time. . . .

Aluminum is still made by the energy-hungry electrolysis technique of Paul Héroult, developed in the 1870s. . . .

The obsolescence of much of this industrial technology is being brought home to Americans by increasing competition from abroad. Europe and Japan, faced with the task of rebuilding their industries after the devastation of World War II, have by now in many cases outstripped the United States in technological innovation. In autos, steel, railroads, and some other industries, the most advanced and productive plants are in Europe or in Japan, not in the United States. . . .

Despite spectacular innovations in some industries, particularly aerospace and electronics, much of U.S. industry, therefore, remains bound to the forms of the first Industrial Revolution.

Given that major bulwarks of the Western industrial economy have been weakened by technological stagnation, the economic giants can only derive small comfort from the fact that the situations in competitive countries are no better or more stable, but are in fact even more worrisome. Innovations are desperately needed at home and abroad.

Once again, as has happened before during the course of economic development since the Industrial Revolution, the countries with the most sophisticated industrial development have built up a crisis potential in their growth industries. Figure 2–8 shows the progression leading to the return of a technological stalemate.

The expiration of the growth cycles in many leading branches of the economy where millions of people are employed is, as we have seen, caused by change in demand and technological inflexibility. Rèpercussions from overexpanded growth in these branches also have a dampening influence on other branches. One can view these restraints on growth as a clogging of many markets with highly standardized goods. To the extent that these goods are mass-produced for the vast majority of the population, they lose their appeal as luxuries and essentials; and people would rather have something else or something different. The reaction against materialism and the existence of a large quantity of standardized consumer goods are two sides of the same coin. They are both causes of stagnation. What are the immediate consequences of these changes in the value system?

The socioeconomic consequences of overexpanded growth in most modern industries were once described as "a caricature of the prosperous society." One consequence is stagflation, which puts the citizens in an industrialized nation into a position where they must worry about the devaluation of their wealth as well as about keeping their jobs. Then there is a transformation in the zeitgeist accompanying the changed relationships created by the stagnation trend; this transformation disturbs the economic atmosphere, thus creating warnings of danger. Let us now examine the sociopolitical effects of stagnation.

STAGNATIONS PRODUCE IDENTITY CRISES

As all things must pass, the only constant factor is change. Habit has made us assume the continued relevance of our past experiences; however, the assault inflicted by innovations and stagnations over time may severely challenge their validity. Experience may be irrelevant; thus, each day erodes our world as a consequence of the interplay between the prevailing socioeconomic structure and the personality development of the individual (Erich Froman in *The Same Society*).

How do modern people react to these assaultive changes that appear as stagnations and invade all aspects of their existence or as innovations that show them new horizons? Do people remain serene? Are they pleased by these exciting opportunities? Or does constant and pervasive change make people feel insecure or even ill?

This is mainly a question of the volume of stagnations and innovations with which a person must cope. What is the general tendency? Do hope and joie de vivre prevail along with pleasure in the rising standard of living that change brings? Or does fear prevail—fear of the dangers that lie behind the fragility and undependability of familiar relationships and objects? My thesis is that the general mood will depend upon the economic expectations, that is, upon whether a climate of stagnation or of innovation is expected to prevail in the next few years.

The year 1968 demonstrates the differences in prevailing outlooks. To gauge the industrial world's mood in 1968, the beginning of the stagnation trend, the French IFOP Institute conducted an opinion pool in eight industrialized nations. Of the Americans polled, 69 percent said that their uneasiness had increased. Only 15 percent indicated that they were reassured by the general turn of events. Of the Dutch polled, 79 percent stated that recent changes had negatively affected their personal situations, whereas only 4 percent believed that their fortunes had improved. Not so for the French! At any rate, 32 percent of the French claimed that they derived a large degree of pleasure from the change; it increased their joie de vivre. Only 27 percent admitted any diminishment in their happiness. Note, however, that it was never determined whether the Americans were particularly irritable because of the Vietnam conflict or whether the French were in fact innately better able to view change in a brighter light.

> The fear is simply there, no matter how well one conceals it behind the protective shield of a perfect living standard.[2]

At any rate, the accelerating forces of socioeconomic change uproot increasing numbers of people and alienate them from their accustomed ways of life. More and more people experience a sense of estrangement in their own socioeconomic context, a fate that Albert Camus vividly portrayed in his novel, *The Strangers*.

Rapid Changes Produce the Impulse Personality

A swift and pervasive change affecting one's entire way of life results in feelings of depression and victimization. This leads to a type of behavior peculiar to alienated people.

> Increasing industrialization and urbanization during the last decades have forced changes in many social structures. Through this progressive change a particular characteristic of static societies, the continuous opportunity for identifications within an undisputed value system, is very often lost. The reactions to this are correspondingly varied. . . . All of these are burdens which the individual has always had to bear; the uniqueness of the situation stems from the combinations and alternating attenuation and intensification of these influences *in rapid succession.*[3]
>
> It begins to dawn on us that the technological world of our day is producing a form of alienation which staggers the imatination.[4]

The modern, estranged human being develops an "impulse personality."[5] The Mitscherlichs have observed and described the characteristic behavior of the impulse personality, beginning with individual details:

> The confusion between such inward and outward experiences is reflected in a contradictory sequence of fragmented behavior; instead of a character with normally predictable behavior, as had been the norm during the bourgeois epoch, one often observes unconnected juxtaposed and highly contradictory sequences of attitudes and actions.[6]

The Mitscherlichs observation of the group behavior of impulsive personalities is particularly significant for our further discussion:

> Paradoxically, the predictability of the behavior of large groups has risen. This apparent contradiction is resolved, however, when we recall that impulsive personalities are particularly vulnerable to the influence of the opinions, attitudes and prejudices to which the mass media expose them. As communication networks connecting the masses become denser, the precision with which it is possible to control the behavior of those dependent upon or accustomed to mass media increases.[7]

Identity Crisis and the Need for New Identity

Socioeconomic change besets people simultaneously with fear and hope. On the one hand, the numerous real stagnations threaten or even ensure their expulsion from traditional, familiar domains in their working world and daily life. "And He drove man out . . ."

(Genesis 3:24). But on the other hand, people are also simultaneously being steered toward new forms of activity in their work and daily life; the abundance of real innovations attracts and stimulates their lust for living and their desire for achievement with a categorical imperative: "Trade ye herewith!" (Luke 19:13).

Present and impending stagnations loosen the bonds of identity that tie people to the old objects and relationships in their working world and daily life. The present or hoped-for innovations offer new possibilities for identification. This "evolutionary game" draws humankind increasingly under its spell: the "immediate responsiveness" of people (and the computer with the technical help it increasingly offers) enable people to "catch the drift of the future."[8] It is paradoxical that modern planning and decisionmaking techniques strengthen the tendency to make more radical decisions that break with continuity.

The loss of old identities and the overwhelming numbers of new identification possibilities available to modern people form sensitive human beings into impulse personalities. The group behavior and market behavior of impulse personalities are determined, on the one hand, by the barrage of stagnations and, on the other hand, by the onslaught of innovations that face them. As individuals, impulse personalities waver uncertainly because of change; however, depending upon whether the change results from stagnation or from innovation, people will bend one way or another. Every economic climate, from the economic boom in the years of growth to the recession from the technological stalemate, produces prototypical syndromes of identity crisis and a need for new identifications.

Change both drowns and resuscitates humankind. If we want to grasp the causes and effects of the alienation and longing for a secure refuge that modern people experience with particular intensity, we should remember that there is a natural period of rootlessness in human life, that is, the period of growing up. Ericson writes:

> Like the trapeze artist, the young human in the midst of intense agitation must release his secure grip on childhood and search for a sure handhold in adulthood. For a breathless interval, he is suspended between the past and the future, between the reliability of what he must give up and what is waiting to receive him.[9]

Two events are necessarily involved in the process of growing up; one must give up one's old identities and simultaneously identify with people and objects that are unfamiliar.

Identity crises and problems of identification also occur in adults who continually encounter new manifestations of stagnation and new innovative possibilities in their daily lives. The evolutionary forces of change throw people off their accustomed course, thus setting them onto new paths. However, when there are few chances for change because of a lack of innovations, we see the truly tragic side of the technological stalemate. Alienation increases sharply, but there are no meaningful alternatives available; therefore people, toolmaking animals, are trapped in a vicious circle:

> The search for roots has continually taken a new form during the mercantile and industrial eras; [look, for example at] . . . the attempt by industrial man to identify himself with the machine, as if it were a new totem animal.[10]

It is not possible for every person set adrift in times of stagnation to acquire a new identity; there are people who have lost their jobs and young people who never have found a decent occupation. The numbers of rootless people seeking refuge vary according to the relative abundance or scarcity of innovations. In a technological stalemate, stagnations outnumber innovations. People in an identity crisis compete with each other for the scarce opportunities to take refuge in innovative areas. Seen from this perspective, the problem of identification for the person uprooted by change becomes an economic problem whose solution must take into account the social and political peculiarities in each group of people.

We will now contrast the situation in the 1960s that arose because of the economic miracle in West Germany with what the technological stalemate later engendered in the 1970s. Basically, it is the same development as in the industrial countries.

The Mute Minorities. We begin with a surge of basic innovations that revives the industrial economy. After World War II, the prewar basic innovations were vigorously exploited. Very few stagnations occurred to dampen the impact of the enormous profusion of the new goods and services and new production plants and techniques. This infusion of new opportunities was one of the preconditions for the economic upswing in all countries in the postwar period; the upturn was further intensified in West Germany by the reconstruction process. Under these conditions, there was a wide field of activity for almost everyone. Jobs were available in all areas of industry and administration, and the hungry markets created a favorable climate for people with business acumen, entrepreneurial drive, and organizational talent.

Manpower scarcity is an economic sign of sufficient possibilities for new identification. Almost everyone under sixty-five who was thrown out of the market by economic stagnations found it relatively easy to find a new job. A mute minority of the sick, the unwilling, and the victims of discrimination, only some of whom had actually been oppressed by economic change, remain as outsiders on the edge of prosperity.

The Silent Majority. After the economic boom is long past (i.e., in our case since the trend shift in 1966–1967), increasingly larger branches of the growth industries as well as the industries that support them have been influenced by spreading stagnations arising mostly from an oversupply of certain goods in the prosperous society. Only a trickle of innovations opposed the swelling stream of stagnation. This imbalance in the relationship between the two evolutionary forces is one of the preconditions for a technological stalemate. In West Germany this imbalance has been intensified by the high dependency on exports and the custom of allowing more kinds of work to be done by foreign workers.

A large number of bankruptcies and the beginnings of massive layoffs on the job market are signs of an identity crisis threatening major segments of the population.[a] The chances for these people to find new occupations are slim because they do not know where to turn. Up to this point, rootlessness and the futile search for other occupations was the destiny of only isolated, alienated minorities. In a technological stalemate, however, it becomes the fate of larger groups and spoils life for the people who are still secure and untouched because they too now fear for their life-styles. They will also have to pay larger shares of the social security costs and other transfer payments to the disadvantaged.

There are few innovations to stave off these crises. Basic innovations have been bogged down in the years of reliance on quantitative growth in existing branches, and if the present tendencies continue, it will take a number of years for any new branches of industry to emerge. In a technological stalemate, it is no longer only a mute minority on the edge of society that is affected by change; the majority of the population is now involved in some fashion even if it is only through an inchoate fear of change occurring. This majority of fearful people is usually silent. However, there are always activists and articulate spokesmen who are able to express the fears of the silent

[a]The identity crisis does not make any exception for professional people, although these groups today really should lie beyond the reach of stagnation. With engineers, for example, the stagnation trend registers as an influence on the "half-life of technical knowledge."[11]

majority. Erikson has pointed out what can easily—and in 1930 in Germany in fact did—happen:

> When historical and technological developments encroach upon deeply rooted or highly upward mobile identities (for example, agrarian, feudal, patriarchal) the youth feels individually and collectively endangered and is, therefore, ready to support doctrines allowing complete submersion in a synthetic identity (extreme nationalism, racism or class consciousness) and a collective condemnation of a completely stereotyped enemy of this new identity. . . . Since the circumstances which undermine a sense of identity also move older people to take up youthful alternatives, a large number of adults march in the same direction or are paralyzed in their opposition.[12]

This phenomenon, of course, is neoconservativism, which may become prevalent. Lack of innovation causes identification problems. Identification with radical movements is an understandable substitution attempt. The purpose of this book is to make the public aware of our scientific evidence that the innovations that can halt the spread of stagnation and offer new identities to members of society have been temporarily held up by circumstantial problems. But basic innovations will eventually appear. The conviction that opportunities for new identities are available in the future could reduce the danger of frustration in the current situation. This optimistic attitude could prevent national governments from considering breaking away from European-Atlantic solidarity and protect the groups in the population now struggling with an identification problem from turning prematurely to radicalism.

We will now depict how the identity crisis and identification problems of the members in a prosperous society have repercussions on the economic process, giving the stagnation trend additional strength. These repercussions also feed stagflation and lead it into slumpflation.

THE STAGFLATION EPIDEMIC AND THE LIMITS OF ECONOMIC POLICY

In Chapter 2, we uncovered the original source of infection that started the stagflation epidemic. We will now show that the worsening of the epidemic occurs through a process of widening the area of infection and of reinfection with the disease.

Stagflation is an artificial word, constructed from the terms stagnation and inflation. It has been coined to convey the concurrence of inflation in the world economy with weakening growth rates in modern industries. In light of the widespread (Keynesian) economic doctrine that inflation is a sign of expansion of economic activity, its exact opposite, stagflation, is clearly a startling phenomenon.

If it were not so difficult for us to relinquish long-held theories when they prove to be inaccurate, the fact borne out by economic history that stagflations have occurred in the past and can therefore occur in the future would have given economic theorists and planners material for thought a long time ago. Although the theory of inflation prevailing today is not wrong and does explain certain circumstances very accurately, it does not apply to today's situation.

The economic process constantly experiences disequilibrium. There are phases of upward development (e.g., after World War II) that begin with relatively severe underemployment; this phase can be overcome by administering a slight increase in the inflation rate. That is the Keynesian message. Then there are phases of downward development in employment (like today) that begin with a relatively high level of employment. The downward trend cannot be halted even with powerful doses of inflation because the downturn is naturally accompanied by stagnation. The popular Keynesian theory of inflation is not universally applicable; in our current situation it is not only impractical, but it is also misleading.

There are many reasons for rising prices; why should the conventional listing of the causes of inflation be considered comprehensive? Prices rise when there is significantly more money and credit available, that is, more buying power than there are goods to sell (liquidity theory or quantity theory). This rough correlation is valid in every circumstance. Keynes's theory of inflationary gaps is both more profound and more limited. When the economy's capacities, inadequate as they are, are fully employed, the producers' first response to the demand surplus is to raise prices in order to raise capital for the expansion of their capacities. A practical application of the theory would involve stimulating investment in order to guide a fully employed economy with low level capacity toward increased growth. The so-called Phillips curve only applies to economic systems with many hungry markets where there is a relatively low level of supply. This is the situation that occurs *after* wars and depressions.

It is a different matter when a prosperous society is fully supplied with all of the usual essential and nonessential standardized goods it could want. Then market satiation creates a classic case of stagnation. This is the phase in the long-term economic process when the economy slides into a technological stalemate and is clearly a very different situation. The huge production capacities for standard goods of daily life can be contrasted to the dwindling demand for these items, and employment becomes structurally unstable.

In this situation, economic theory predicts a deflationary rather than an inflationary trend; historically and theoretically, stagnations are accompanied by sinking prices. Indeed, in 1976 many prices for German industrial goods have fallen under the 1975 level. About one-sixth of the positions in the price index—mainly consumer goods—have negative signs (for example, vegetable oils, 17 percent; bicycle tires, 12 percent; black and white TV, 7 percent; colored TV, 5 percent; cement, hifi sets, and carpets, 3 percent).

During the late stage of stagflation (Phase III) when the price mechanism in sectors of elastic demand has its way, price decreases put pressure on overproduction in these sectors. On the other hand, the population may still suffer from inflation because in sectors of inelastic demand the need to buy even at increasing prices allows for prices to be increased. This is true for a number of household budgetary items such as car insurance fees, health costs, and utilities that the regulated industries provide. And, of course, it is also true for taxes. This is why the late stage of stagflation turns depressive.

In the following paragraphs, I shall focus on Phase I and II of stagflation. Phase I is the period when the causes of both inflation and stagflation become effective and initiate stagflation. Phase II is the period when the effects of stagflation reinforce both stagnation and inflation, and stagnation feeds on inflation and inflation feeds on stagnation. I shall attempt to integrate the self-sustaining stagflation process into a *liquidation theory* that tries to explain the "move from stagflation to slumpflation."

Concentration Causes Stagflation

Phase I in the stagnation period of the maxicycle sees recessive tendencies as breaking the prosperity trend. This was what occurred in the 1966–1967 slump. While demand for most products and services of luxury life and daily use was still strong, structural change in supply created trouble. Industrial concentration—more mass production for the sake of returns to scale, on one side, and more bankruptcies and insolvencies of smaller and medium-sized firms, where in European countries more than two-thirds of the industrial labor force is employed, on the other side—resulted in unemployment problems that from the mid-1960s on were no longer resolved by short-term booms.

The Economist (of October 9, 1976) reported a study of industrial concentration in Great Britain, the results of which must be considered a conservative estimate because in Britain "progress towards industrial concentration was, in the twentieth century, slower than in competitors like America, Germany and Japan." Over the years 1958, 1963, and 1968, in Britain the average concentration ratio of the five largest companies in each of the 198 product markets examined rose from 55.5 percent to 64 percent during this decade—"a pretty dramatic increase" (*The Economist*). Table 3–1 shows the extent of industrial concentration according to the 1971 census in Britain.

Our thesis is that concentration increases the market power of suppliers and allows them to supply less than is best to the consumer, to slow down the rate of quality improvement, and also to charge higher prices than are attainable under conditions of more quality and price competition. Studying a cross-section of thirty industrial branches in Germany in 1974, Herbert Schui found that the higher the concentration ratio (share of the six largest firms in the sales volume on a market), the higher the inflation rate and the lower the growth rate in that branch (Table 3–2).

Table 3–1. Concentration in British Manufacturing Industry, 1971.

Order	*Products*	*Number of Products Analyzed*	*Concentration of Sales Among 5 Largest Enterprises*[a]	*Total Sales of Products Analyzed £m*
II	Mining and quarrying (except coal)	2	50	77
III	Food, drink, tobacco	12	57	3,916
IV	Coal and petrol products	1	36	44
V	Chemicals	14	64	2,308
VI	Metal manufacture	3	61	647
VII	Mechanical engineering	13	41	1,172
VIII	Instrument engineering	3	46	211
IX	Electrical engineering	8	63	2,069
X	Shipbuilding and marine engineering	1	47	311
XI	Vehicles	4	73	3,491
XII	Other metal goods	8	49	675
XIII	Textiles	12	43	950
XIV	Leather, leather goods, fur	3	24	55
XV	Clothing, footwear	8	27	334
XVI	Bricks, pottery, glass, cement	5	53	604
XVII	Timber, furniture	6	16	208
XVIII	Paper, printing, publishing	8	43	1,096
XIX	Other manufacturing	4	37	57
	Total	115	47	18,225

[a] Weighted by total sales in products covered.

Table 3–2. Stagflation and Concentration.

Annual growth rates and price change rates in three groups of industries classified by low, medium, and high degree of concentration

	Group 1	*Group 2*	*Group 3*
Number of branches in the three groups	6	15	9
Average concentration ratio in each group	9.3	25.6	66.2
Average price change in the respective markets	6.9	9.4	11.4
Average production change in the respective markets	–4.3	–8.6	–9.5

Source: Wirtschaftswoche, No. 13, March 18, 1977.

In the United States, the stagflationary consequences of industrial concentration were further strengthened by the fact that concentration occurred more often than in other countries in the form of conglomerate mergers. These mergers add more to market power than simple horizontal or vertical integration. Figure 3–1 shows that in the late 1960s, the United States experienced "the third great merger movement in American history."

My institute colleague, Dennis C. Mueller, in a survey of the empirical evidence on the effects of conglomerate mergers,[13] describes the first major merger wave as a wave to create monopolies, the second to create oligopolies, and the third to create conglomerates. The distinguishing feature of the mergers in the 1960s was certainly the extent to which these mergers tended to diversify or extend the acquiring companies' product mixes. The empirical evidence collected by Mueller suggests that of the two obvious reasons for merging, higher profits or more market power, the first clearly did not materialize although the second did. Because a conglomerate has both more freedom to gamble on higher prices in specific markets and to offset the losses from risky investments, the conglomerate merger wave must be seen as an amplifier of both inflation and stagnation. The biggest and fastest growing conglomerate is the government, and it behaves as a conglomerate.

It is thus more than random coincidence that the stagflation trend of the 1920s and the stagflation trend of the 1960s were both accompanied by merger waves. Figure 3–2 shows this with an index of merger activity (1919–1920 = 100) based on a number of mergers between 1895 and 1972, thus adding further evidence to our metamorphosis model (Chapter 2).

Figure 3–1. The Recent Conglomerate Merger Wave.

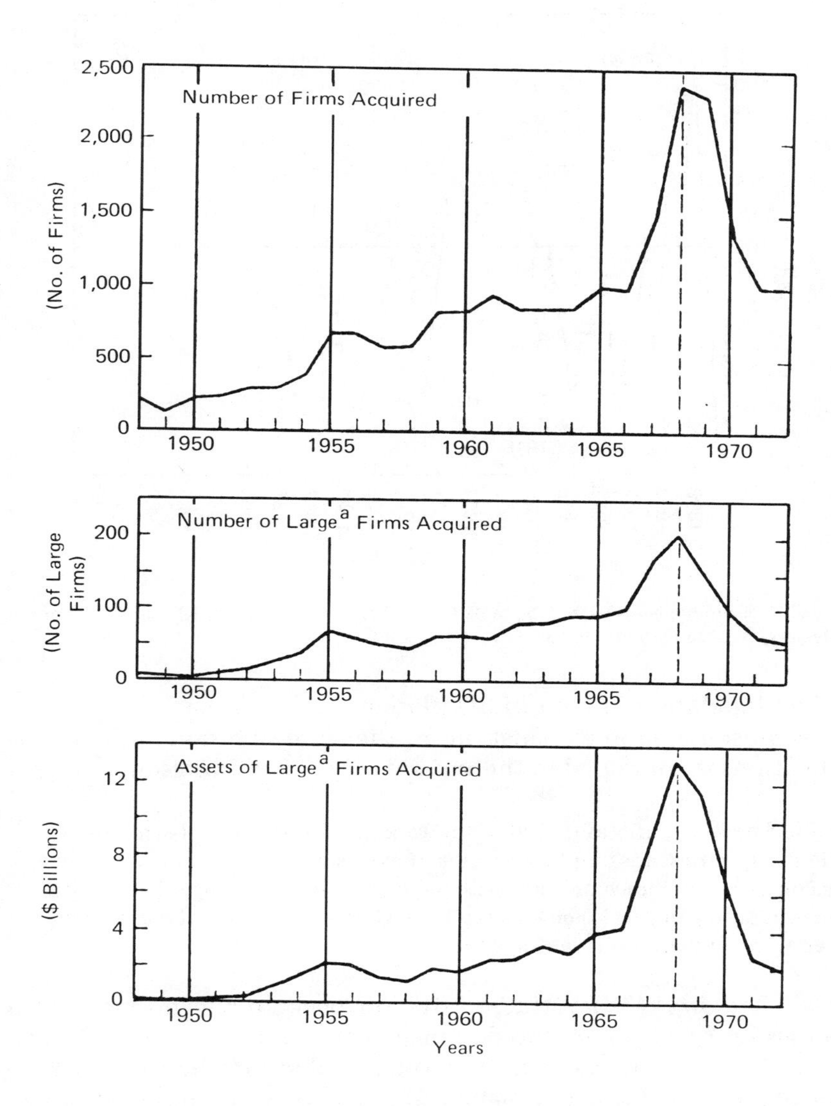

[a]Firms with assets of $10 million or more.

Source: Federal Trade Commission, Bureau of Economics, *Current Trends in Merger Activity, 1971,* Statistical Report No. 10 (Washington, D.C., May 1972), p. 6, and *Report on Mergers and Acquisitions* (Washington, D.C., October 1973), p. 147.

Figure 3–2. Merger Waves in Times of Stagnation.

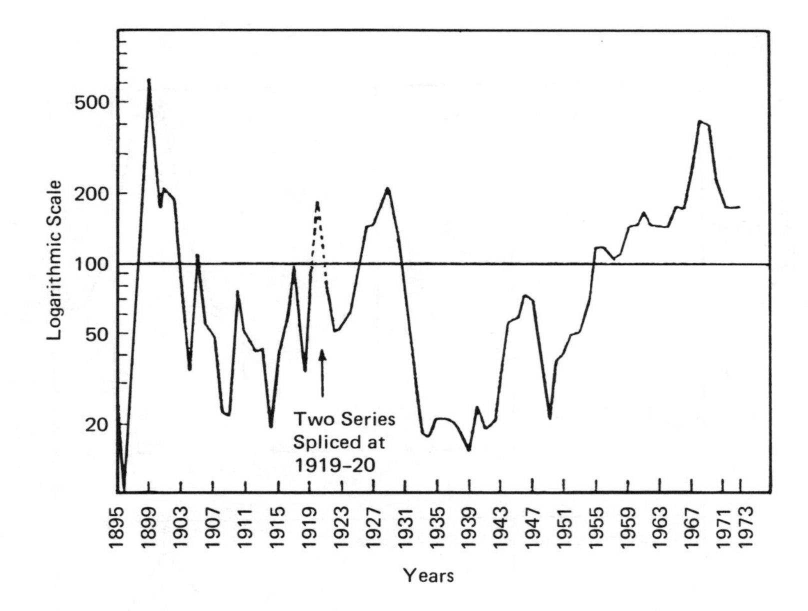

Sources: Nelson, *Merger Movements*, table C-7; FTC, *Current Trends*, 1971; FTC, *Report on Mergers and Acquisitions*, 1973.

The Liquidation Theory of Stagflation

The most common explanation for the price situation in the industrial nations is contained in the liquidity theory of inflation:

> Incidences of inflation, despite the variations in shape and size for individual countries, are simply the result of excessive liquidity which does not accumulate by choice or unexpectedly, but—"and here Edgar Salin points an accusatory finger," is engineered through the "wisdom" of currency experts' and experts from central banks.[14]

This *liquidity theory* traces inflation back to the mistakes of a small group of planners who do indeed bear a large share of responsibility. However, in the above quotation they are being forced to shoulder too much of the blame. The exclusive emphasis on monetary problems does not do justice to the complexity and power of the real-economic tendencies—concentration in supply and satiation in demand in many markets for industrial goods and services—that lead to stagflation. At this point, I will attempt to fill in some of the gaps in the liquidity theory.

Stagflation is set into motion (Phase I) when the concentrated network of suppliers eliminates quality competition from the major market segments supplying increasingly fewer essential standardized goods to largely satiated consumers. Pseudo-innovations indicate that the pressure to keep costs and prices down is eliminated. For a certain period, stagnation and unrestrained price increases run in tandem (Phase II).

Once set into motion, the stagflation disease is transmitted via financial and marketing mechanisms. This process lives off the energy generated by structural decomposition. Stagnation of demand in leading sectors is communicated to the supplying industries and also affects the turnover in the capital goods branches. Inflation of cost components (wages, raw materials, and working capital) puts pressure on profits. Inflation intensifies stagnation. This process is depicted in Figure 3−3.

The liquidation theory deals with the feedback of stagnation on inflation. A fundamental change in the socioeconomic climate lies at the root of the connection between stagnation and useful innovation. Stagflation begins when the favorable ratio of few stagnations to many innovations changes to an unfavorable ratio of many stagnations to few innovations. At this point the identity crisis that is caused by these stagnations shifts from being a problem for a minority of the population to a concern for the majority. This identity crisis is like a prairie fire burning in a drought; there are very few identification opportunities available to douse the fire. Scarcity of innovations means that there are no hopeful alternatives for the people threatened by stagnations.

The circle of stagflation is continually enlarging; as the minority threatened with expulsion from their economic positions grows, it includes more groups of employees and business people who have only a short time left in their familiar situations before they must yield to structural change. Yet these groups can still offer resistance. How?

Threatened by the surge of stagnation and yet still present in the marketplace and therefore able both to influence price levels and to demand certain salaries, business people and other occupational groups initially react in the same manner. Since no other solution is in sight, they take refuge in demanding higher salaries and prices. "They want to make money while the sun shines," a character in *Sam* explains. *Sam* is the story of an English mining family in the 1930s.

Like the English miners during the structural crises of 1930 and 1974, the farmers plagued with overproduction continue to demand

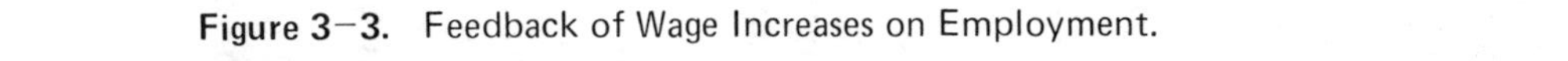

Figure 3–3. Feedback of Wage Increases on Employment.

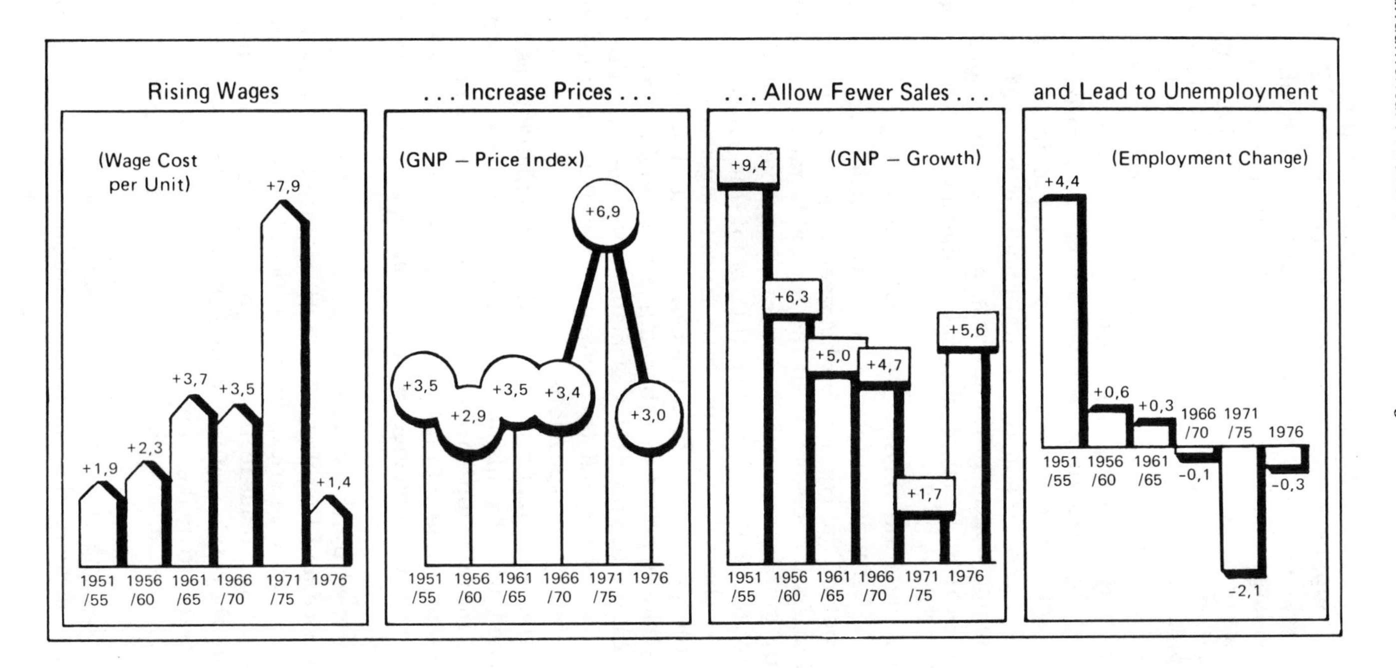

higher prices for their products as long as they are still being purchased. Farmers and miners are often front-runners in the race for higher prices and wages because both occupational groups are well protected from the consequences of a full-scale wage and price war by a system of government subsidies, guarantees, and unemployment insurance.

Other occupational groups behave more moderately because they know that the system of unemployment insurance will collapse if the strain becomes too great. In any event, the modern welfare state is completely controlled by interest groups; in most countries the producers in particular have closely allied their associations (employer associations and labor unions) with major political parties. At the opposite pole, consumers have no effective pressure groups and absolutely no dependable political support. The consumer ends up paying for everything.

Middle-sized businesses also tend generally to lack protection from the effects of stagnation and must close down when the market pressure mounts. The number of bankruptcies in West Germany rose from 2,239 in 1966 to 5,277 in 1973 and over 10,000 in 1976. This trend correlates well both with the increasing inflation over the same period and with industrial concentration.

Even the highly criticized oil companies whose profiteering during the oil crisis excited worldwide indignation have provided support for the liquidation theory in their recent actions. The nationalization or partial nationalization of the oil fields in the producer countries has robbed these companies of their strongest source of profits over the medium term. The oil companies used to make their biggest profits at the source of the oil; in the marketing network, profits are much lower. Since they now can only expect to make profits at the marketing stage, they have no other recourse than to raise prices.

Thus, stagnation encourages price increases. Using the liquidation theory, we have explained the behavior of those caught in the pinch of the structural crisis. Other economists have also observed and evaluated this behavior. In fact, Giscard d'Estaing, the President of France, wrote an article discussing four kinds of inflation in which he stated:

> It must be acknowledged that to fight this inflation, which is the most worrisome type in modern countries, economic thinking is no more advanced than [current] medical thinking is about ways to stop the spread of cancer cells." . . . I am convinced that, one way or another, structural changes . . . are reflected in inflationary tensions. This involves phenomena such as migration from farms, [etc.]. . . . It seems more and more that the

> connection observed between unemployment and inflation basically stems from the anticipatory behavior of consumers and workers, that is, short-term adjustments.[15]

If he had only included the short-term reactions of producers and merchants, Giscard d'Estaing would have produced a formulation of the liquidation theory.

Two political economists from Oxford have also established a connection between stagnation in industries and the fate of the entrepreneur in the structural crisis. They have shown that the British economy since 1964 has become progressively less profitable,[16] and that since the mid-1960s, with the increased flood of stagnation, its short-term indebtedness has skyrocketed while the amount of available capital has sharply decreased. What is the result? Twice as many bankruptcies.[17] And what is businessmen's first reaction to these critical developments? According to Glyn and Sutcliffe: "Putting up prices."

This then is the liquidation theory of stagflation. It provides a real-economic explanation that states that people who see their socio-economic positions threatened by stagnations extricate themselves from these untenable relationships and liquidate their holdings. The impulsive personalities then turn from the long-term cultivation of their sources of income to short-term exploitation. The gardener becomes a hunter—a price hunter. Because expulsion from traditional socioeconomic positions and the fear that this instills in observers who are similarly situated have affected progressively larger groups of employees and employers during the late phases of the growth cycle, identity crises are also widespread. These crises lead people to adopt the attitude that they will simply snatch at everything that the market has to offer before its collapse.

The following two sectoral (as opposed to global) inflationary factors therefore can occur in conjunction with each other:

- The liquidation theory of stagflation explains price increases as a short-term rescue operation by people who are threatened by the structural crisis; the forces of stagnation are pulling them away from their places of employment, their farms, their businesses, or their markets.
- In contrast, the Keynesian theory of inflationary gaps explains price hikes as a short-term exploitation of favorable market conditions in which the excess demand allows workers and entrepreneurs to ask for more money for the same amount of work.

Clearly, both of the tendencies pushing prices upward can exist simultaneously in a pluralistic market economy; in fact one can even feed off the other (e.g., a boom for pawnshops and secondhand stores). However, when as today many markets have been drawn into the morass of stagnation and only a few are prospering, there is a marked tendency to turn to pricing policy as a short-term panacea. In any event, monetary policy faces a dilemma.

Most Keynesians view wage and price increases in flourishing markets experiencing excess demand (seller's markets) as the most powerful inflationary forces. Therefore, these theorists' predictions that the inflationary tendency must level off during stagnation proved inaccurate. They overlooked the fact that with increasing stagnation and as a result of business liquidations the short-term tendency to increase prices became overwhelming. This is because the vast majority of the small and middle-sized businesses are oriented to a buyer's market. This means that economic planners who wish to take measures to revitalize the economy or stabilize prices, thereby either expanding or restricting the money supply and the credit available, end up dampening one inflationary tendency but simultaneously encouraging the other.

In 1973, during a time of increasing stagnation and rising prices, governments in many Western countries implemented an anti-inflationary policy with money and credit restrictions. This appeared sensible in light of both the liquidity theory and the Keynesian theory of inflationary gaps. According to the liquidation theory, however, it was a mistake. The tight money policy in fact intensified the tendency to increase prices and entangled more businesses in the crisis. By encouraging a slowdown (stagnation), the unfortunate anti-inflationary policy intensifies both stagnation and inflation. This is exactly the misfortune that occurred in the United States with Ford's "Fight against Inflation."

This misfortune has a parallel in the developments during the 1920s, a decade that evolved structurally similar to the past ten years. Then as today inflation was the major concern for both citizens and economic planners; many Western nations therefore took refuge in deflationary policies. In short order, they achieved exactly the opposite of what they desired. Some years ago, Friedman and Schwartz astounded their colleagues with the assertion that in actuality there could have been no inflation whatsoever during the 1920s. During this period in the United States the amount of money available apparently did not exceed the market need for money. Therefore, they say, according to the liquidity theory there was no room for inflationary tendencies to develop.[18] Kindelberger recognized the same contradiction in contemporary situations in Germany and England: "The German government followed a policy of deflation, just as the English did. Both were fighting an inflation which did not exist."[19]

What can these confusing statements by our leading economists teach us if the fact is that stagflation was a pervasive phenomenon in the leading industrial nations before the world economic crisis of 1929? The famous explosion in prices and the concurrent stagnations during the 1920s communicate an unmistakable message. The helplessness experienced by economists dealing with stagflation cannot be justified by the novelty of this phenomenon. Therefore, the suggestion of the monetary theorists that no inflation could have occurred because the limited supply of money and credit could not have sparked inflation clearly involves accepting conjectures from a theory as facts.

Actually the reverse situation conforms to reality. During the period of stagnation, despite the deflationary policies that imposed limits on money and credit, a whirlwind price spiral developed. This provides a classic example of the flight into higher prices and wages with which employees and employers in the structural crisis-ridden areas of the economy abandoned hope for the sluggish markets. The fact that stagflation occurred in the 1920s as well as in the recent past establishes that neither the liquidation theory nor the theory of inflationary gaps is universally applicable; the first holds for times of stagnation, the second for times of recovery.

In intensifying stagnation, policy guided by the wrong theory only makes matters worse. A deflationary money and credit policy only heats up the inflation over the short run. Over the medium term it can lead to a brutal economic collapse similar to 1929 because this policy intensifies stagnation. A deflationary monetary policy only encourages stagflation over the short and medium terms, and if stagflation progresses too far, it can break the backbone of the business world. The collapse in 1929 conceivably could have been avoided were it not for this "suicidal policy" (W.S. Woytinsky's expression for the highly restrictive measures taken by the German Chancellor Henrich Brüning, 1930–1932).[20]

More recently, however, one of the U.S. presidents played a dangerous game. If his restrictive policies were not followed, warned President Ford in a speech in Oregon on November 2, 1974, the inflation rate in America could double by 1976. How could Ford's advisers suggest such a foolhardy "fight against inflation"? Surely, the similarity of the economic situation at that time to the situation prevailing during the late 1920s should in itself provide an impressive warning against the thoughtless use of a restrictive policy. The West German Chancellor Helmut Schmidt perceived the parallel, and Giscard d'Estaing appealed for moderation with this warning: "Some countries have gone through the sad experience of situations in which rising unemployment was in no way accompanied by declining inflation: stagflation thus becomes a major risk with anti-inflation policies which rely too heavily on cutting back economic activity.[21]

In the 1920s "one discovers that there were deflationists everywhere—Hoover, Brüning, Snowden, and Laval."[22] The same holds true for the early 1970s. However, the suicidal deflation policy of the 1920s, which "in hindsight appears as economic illiteracy,"[23] had some justification at that time given the prevailing belief that there was nothing else that could be done. Treviranus, one of Brüning's cabinet ministers, showed his frustration at this apparent lack of alternatives with a Cockney explosion: "Blimey, there's nothin' else to do!"[24] Heinrich Dräger, the well-known Lübecker industrialist who during the 1920s saw that expansionist policy could provide a way out of the crisis, deplored the credibility given to deflationary policies by the leading economists. His suggestions for using government contracts and easier credit to create new jobs were characterized "with great solidarity by numerous professors on various economic faculties in Germany as unworkable, impractical, inappropriate or at the very least, irrelevent."[25]

Today, on the other hand, most leading economists in almost all of the industrialized nations are Keynesians; many of them put great faith in the effectiveness of an expansionary policy—and this faith is clearly spreading. President Ford depended upon the effectiveness of an expansionary policy so absolutely that he dared to flirt with economic collapse in his deflationary fight against inflation. It was as if one really knew the limits to the economy's tolerance and could quickly reverse a slump with an expansionary policy as if it were a safety net. The largest unanswered question today is whether our experience with expansionary policies justifies faith in its effectiveness?

In Germany, an expansionary policy has been consciously implemented only three times. The first time was in the 1930s (the famous-infamous Labor Front involving highway construction, and so on); the second time was after World War II (the Marshall Plan funds, government credits, investment tax credits, and so on). Both times, however, the prevailing conditions were very different from today's. During these periods, Germany was suffering from severe unemployment—in the first instance because of the world economic crisis and in the second because of the war. This meant that Germany, far from being a satiated prosperous society, was a country with an undercapitalized economy facing real privations. The conditions under which the effects of an expansionary policy would have been of interest to us were the conditions existing before, not after, the world economic crisis. During the golden 1920s, Germany, like its Western neighbors, was a prosperous country; the majority of its population was fully supplied with the products that it needed or desired. At that time, however, there was no move toward an expansionary policy; in fact, as we have noted, many experts maintained that such a move would be senseless.

However, does the third instance not meet our qualifications? Does the expansionary policy during 1966–1968 not provide an encouraging example? The situation in 1966 certainly provides a closer parallel to today's conditions than the first two examples do, not simply because of its historical proximity, but also because of the similar levels of economic activity. In 1934 and 1949, the unemployment rate was very high; compared to these periods, 1966 and 1967 involved much lower levels of joblessness.

The essential difference between the situation in 1966 and today can be seen in the relationships between productive capacity and demand. In 1966, the productive capacity of many markets was large but not overexpanded, while demand had fallen off but had not become completely stultified. Today, on the other hand, suppliers are dealing with overgrown production capacities, and goods are not moving swiftly out of stores or inventories because demand has dwindled down with the market no longer absorbing the mass of mass-produced goods. In 1966 there was still room for expansion of existing capacities on a modest scale if some competitors could be thrown out of competition at the same time. Obviously, monetary measures cannot mobilize production and demand if there are non-monetary, structural obstacles, namely, discrepancies between the production and need structures.

In the last twelve years, the consumer sentiment index (Figure 3–4) indicates a fundamental change in the demand schedule and in people's needs. The change in the willingness to buy what is offered cannot have been caused by monetary factors like lack of disposable income or buying power alone. It would be hopeless to try to cure stagnation purely by monetary policy measures. Our proposition, namely, lack of demand-inducing innovations has caused the stagnation of final demand, has some support from Peter Temin's analysis. Recently, Temin concluded from his noteworthy study of the slumpflation in the 1920s: "The proposition that monetary forces

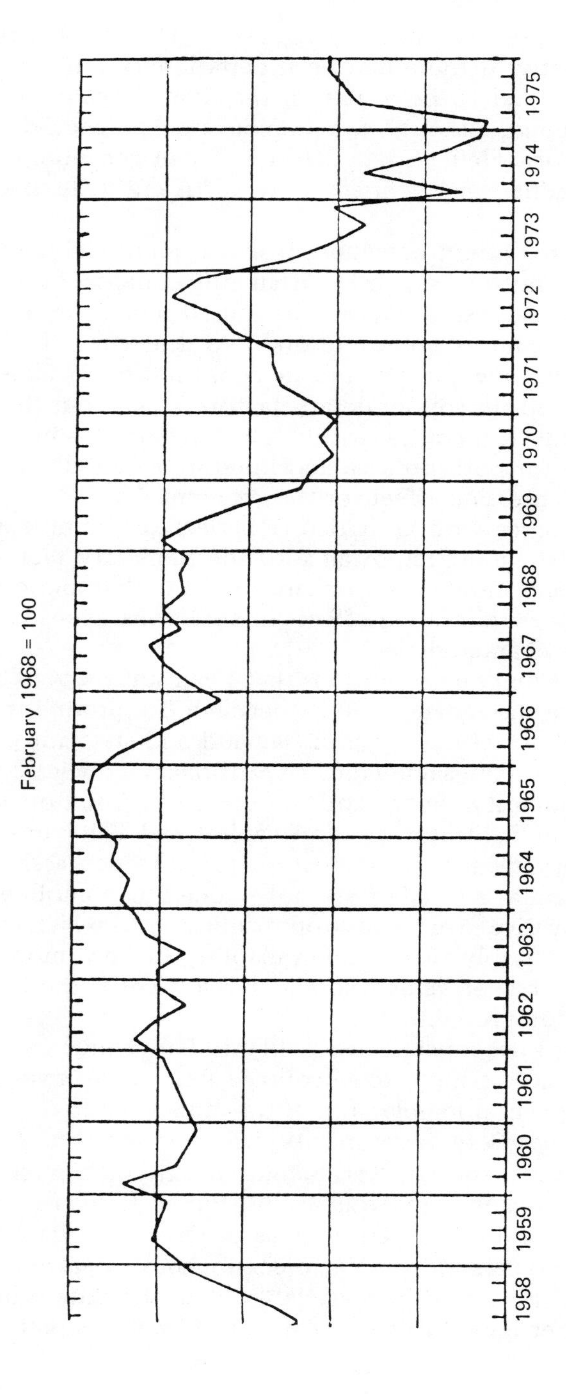

Figure 3–4. Michigan Index of Consumer Sentiment in the United States.

caused the Depression must be rejected." He believed this for two reasons: "Firstly, if there had been deflationary monetary pressure, it would have had to be visible in the financial markets," which it did not. "Second, although the nominal stock of money fell in 1930 and 1931, prices fell also. . . . If the fall in the nominal stock of money was deflationary, prices were sufficiently flexible to absorb this pressure."[26]

Furthermore, Temin concludes from the empirical facts, "The Depression was not caused by a dramatic collapse of investment." Rather, as is the case today, it was caused mainly by nonmonetary factors that reduced consumption: "At the current state of our knowledge, the unexplained fall in consumption is larger than the part we can explain (by monetary factors, G.M.), but the magnitude of the total fall is incontravertible. The large decline in consumption expenditures for both durable and nondurable goods in 1930 had a profoundly depressing effect for the economy."[27]

Thus, I disagree with Gottfried Haberler, for example, who attests to "the overwhelming importance of the monetary factor."[28] However, with many monetarists I share the view that monetary policies can have little equilibrating effects in situations of disequilibrium in nonmonetary characteristics.

Given the economic conditions that I call stalemate in technology, that is, where discrepancies exist between the production structure and the need structure, sluggish demand and stagnant employment are two sides of the same coin. Expansionary policies based on the wisdom of liquidity theory and the theory of inflationary gaps will not succeed in fighting structural capital and labor unemployment with monetary means. As stagnation turns to depression (Phase II, Phase III), capital and labor are not scarce but immobile. There are too few innovative investment opportunities into which capital could safely and profitably flow. The availability of new, more useful, less wasteful, and higher valued goods and services is the bottleneck in this reallocation problem.

In addition, an expansionist policy today cannot count on being successful if it attempts to encourage innovative investment using the watering-can principle; the extra money poured into areas already suffering from overcapacity will not produce any capacity-increasing investment. (As Karl Schiller put it you can lead a horse to water but you cannot make it drink.) Public funds can be effective only when they are invested in areas where the limits to growth do not exert a braking effect. Yet precisely these areas attract the private capital that has become available. Thus the state is merely pouring more water into the ocean. This wasted water is precious because there are areas where private capital cannot or will not flow, areas of potential basic innovations. The infusion of public funds here would make sense—and if successful—would create a multiplier effect, that is, new jobs and new demand, and more jobs and more demand. Innovative investments are the stimulants of qualitative growth.

The starting point for crisis management lies in corporate strategies and public policies for innovation. Improvement innovations in established industrial branches keep the dwindling disparities in technological quality from disappearing and allow a competitive advantage in national and international trade even under the difficult conditions of a technological stalemate. Because of this innovative edge West Germany experienced less stagnation than most other industrialized nations in the last few years. Germany has therefore also had relatively low rates of bankruptcy and unemployment and has not had to resort to price increases to the same extent that other countries have. The rate at which prices have risen in Germany was and is lower than that for most other Western nations. More innovation means less stagnation, less inflation, less stagflation, less liquidation of businesses, and less unemployment.

The only lasting cure for the economic crisis lies in basic innovations. Although there are large-scale projects that require the help of the government's organizational and risk capital distribution facilities, the vast majority of innovations and contributions to human welfare come, not from the bureaucrats and functionaries in the government and private economy, but from bold-spirited people with energy and initiative who do not lose courage when times are rough. These people give their best to their work: imagination, initiative, talent, persistence, and concern for the needs of others, which is the major factor of success in innovation.

SUMMARY

Stagnations have their roots in the exhaustion of the possibilities for improvements in old technologies, which then facilitates the concentration of supply and a satiation of demand. Both of these phenomena are an expression of the prosperous citizen's new unwillingness to buy standardized mass-produced goods in the huge quantities in which they can be produced by the greatly expanded growth industries.

The people whose economic lives are unsettled by the changes in circumstances (stagnation = identity crisis) search for alternatives in other areas of activity (innovations = possibilities for new identifications). Initially in a technological stalemate there are few alternatives at hand (identification crisis). The result is an exploitation of the now threatened positions as long as this is possible (liquidation), occurring over the short term through resort to higher prices and wages. All of this microeconomic behavior is detrimental to the macroeconomic stability.

Many Keynesian economic theorists and monitarists misunderstand stagflation. They assume that inflation and stagnation are symptoms of two different kinds of situations. In reality, price increases are caused in part by the liquidation of economic positions expected to be untenable (the liquidation theory of stagflation). To a large degree, price increases are a symptom of the swelling number of difficulties in a stagnant economy.

Monetary and credit policies are powerless here because they cannot differentiate between stagnating and flourishing markets. A tight money policy cannot hold inflation in check since it also increases the flight into higher prices and an easy money policy cannot stimulate investment when the lack of workable innovation projects prevents the capital owner from investing in innovation. There is surely a "need for a wholesale reconstruction of the micro-foundations of contemporary monetary theory."[29]

The starting point for crisis management lies with useful projects. Only innovative projects can provide worthwhile investment opportunities. This is why an expansionary policy will be unsuccessful if the public funds are distributed according to the watering-can principle; the money will simply dissipate without an expansionary effect. Ideally, public funds should not be used to produce what would be produced even without their help; they should be used to make what is now only feasible into a reality.

REFERENCES

1. The North Pole Papers, *Environment*, 15 (1973), pp. 2–38.
2. Friedrich Sieburg, *Die Lust am Untergana* (Hamburg: 1961), p. 44.
3. A. und M. Mitscherlich, *Die Unfähigkeit zu trauern* (Munich: 1967), pp. 225–226.
4. E.H. Erikson, *Einsicht und Verantwortung* (Frankfurt/Main: 1971), p. 91.
5. A. Mitscherlich, *Auf dem Wege zur vaterlosen Gesellschaft* (Munich: 1963), p. 280.
6. A. und M. Mitscherlich, *Die Unfähigkeit zu trauern*, p. 226.
7. Op. cit.
8. P. Teilhard de Chardin, *Die lebendige Macht der Evolution* (Alten-Breiburg: 1967), p. 244.
9. Erikson, op. cit., p. 77.

10. Op. cit., p. 92.
11. *Der Spiegel*, 1.5 1972, p. 62.
12. Erikson, op. cit., p. 80.
13. Dennis C. Mueller, "Survey of the Empirical Evidence on The Effects of Conglomerate Mergers," *IIM*, Berlin (1977).
14. E. Salin, "Die EWG in Koma," *Kyklos*, 26 (1973), p. 732.
15. V. Giscard d'Estaing, "Four Types of Inflation," *Economic Impact* (1974) 7, pp. 18–22.
16. A. Glyn and B. Sutcliffe, *British Capitalism, Workers and the Profits Squeeze* (Harmondsworth, 1972), p. 66.
17. Op. cit., p. 126.
18. M. Friedman and A.J. Schwartz, *A Monetary History of the United States 1867–1960* (Princeton: Princeton University Press, 1963), p. 298.
19. Ch. P. Kindelberger, *Die Weltwirtschaftskrise 1929–1939* (Munich: 1973), p. 173.
20. W.S. Woytinsky, *Stormy Passage, A Personal History through Two Russian Revolutions to Democracy and Freedom, 1905–1960* (New York: 1961), p. 466.
21. V. Giscard d'Estaing, op. cit., p. 22.
22. Ch. P. Kindelberger, op. cit., p. 21.
23. Op. cit.
24. G.R. Treviranus, "Das Ende von Weimar," in *Heinrich Brüning und seine Zeit* (Düsseldorf: 1968), p. 1973.
25. W. Grotkopp, *Die Große Krise* (Düsseldorf: 1954), p. 38.
26. Peter Temin, *Did Monetary Forces Cause the Great Depression* (New York: Norton, 1976).
27. Temin, op. cit., pp. 170–172.
28. Gottfried Haberler, "The World Economy, Money and The Great Depression" 1919–39, 1976.
29. R.W. Clower and P.W. Howitt, *Journal of Political Economy*, 1978, No. 3, p. 449.

Part II

The Transfer of Knowledge

"Thought is the seed of action" (Emerson)

It will be possible to stave off the dangers of stagnation that are caused by discrepancies between the supply of private and public goods and the true needs of the population only by encouraging qualitative growth, that is, structural change via new technologies and social innovations that serve actual needs:

- It is the companies, production and trade operations, and individual branches of the economy that experience these discrepancies as a reduction in the volume of orders received; this stagnation causes insecurity in both employers and employees because of its threat to their sources of incomes.
- It is the cities in which most of the industrial workers live where stagnation as a reduction in workers' incomes also affects the well-being of the middle-class, who relies on the workers' demand for their services (from building superintendents to doctors to barkeeps). Ultimately, this also serves to reduce the government's spending power.

Because stagnation in certain areas eventually overflows into other areas and thereby affects more individuals, it should be viewed as a community problem and not shunted off as a peculiar concern of only some groups. The solution to this structural problem will require an adequate number of technological and social innovations in both the private and public sectors of the economy. The process of implementing innovations is common to structural changes both in the regional and sectoral fabric of the economy. In urban dynamics as in industrial dynamics, the establishment of new industrial enterprises and public or semipublic service sectors is the critical variable. It provides the initial impetus that produces shifts over time in the industrial branch and regional structure of the economy. This process of structural adaptation to changing needs results in both progress and prosperity. However, in a technological stalemate, there is a lack of innovation.

Economic and social changes in the working world and daily life occur in an evolutionary interplay between stagnation (in withering branches of the economy, neglected areas of a city, or eroded tracts of land) and innovations (in flourishing branches of the economy, attractively designed areas of a city, and a well-cared for environment). Stagnation is the expression of a lack of improvement innovations in the areas in which it has cropped up; it is often due either to a rigidity in the institutional order blocking any possibility of change or to a dearth of new ideas. If some skills and crafts become obsolete; others can and must take their place to compensate for the loss. This is why basic innovations are necessary. They must be invented, however, and thus the technological stalemate signals a heyday of creativity.

The perception that research findings, seemingly wholly of academic interest, could somehow have a practical application and the hope that society might possibly benefit from these achievements are the thoughts that keep the inventive spirits working, even when this creates a barrier of misunderstanding and dissonance between their contemporaries and themselves. The sources for creative mutations lie in this esoteric research and garret art. Diverging ideas are transmitted in the knowledge transfer, and when they survive, it is as technical or nontechnical basic innovations and improvement innovations.

During the industrialization, applicable information has become a decisive factor in the setting-up and functioning of production and service delivery systems. My empirical analysis of the knowledge transfer process reveals: philosophers of science do well in cooperating with sociologists of science if they analyse the birth and spread of ideas and knowledge; but if they are concerned about implementation, they better cooperate with economists and management scientists. *Application is economics.*

The cultivation of this idea perception process is what allows new ideas to be born and what keeps our socioeconomic life going over the long run. The reasonable person provides for the future. Control of this perception process and of the transfer of knowledge from theory to practice is as important for the human society, now that it has passed its archaic phase, as are the control of the birth process and the socialization of the new generations. It is important that neither too few nor too many innovations be created.

However, precisely in the alternation between too few and too many basic innovations does the destabilizing power of "the process of creative destruction" (Joseph A. Schumpeter) reveal itself over time. If the slide into a technological stalemate can be attributed to the slackening of the innovation rate, then the pressure generated by the poverty of the unemployed and the increasing burdens on the employed may very well pave the way for another surge of basic innovations in the future. However, not all of the basic innovations that might emerge under these unfavorable circumstances are in fact desirable innovations. The pressure of unemployment can easily engender industrial innovations, such as super yield weapons, gargantuan urban renewal projects, gene mutation, hydralike highway systems, and atomic waste disposal technology whose later effects may produce substantially more harm than good. This Faustian future has now become a possibility with which we must reckon; this is the real challenge of our time and civilization. In fact, it is the essence of the normative problem in the present technological stalemate.

In the following three chapters we will investigate the fluctuations between scarcities and abundance of basic innovations, deal with the question of the origins of these surges of innovations, and show with what vehemence these spurts occurred during the technological stalemates of the past. This alternation between innovative scarcities and surges of basic innovations naturally generates expectations about the next surge of innovations; a trend extrapolation based upon our metamorphosis model (developed in Chapter 2) will be given in Chapter 7.

 *Chapter 4*

The Ebb and Flow of Basic Innovations

"What is now proved was once
only imagin'd" (Blake, Proverbs of Hell)

Stagnation means lagging or even negative growth. By standing still one can be losing ground. Zero growth as well as adjustments from quantitative to qualitative growth produce management problems; if they are not resolved, the economy will remain stranded in the limits of growth within the cage of its present structure.

In a recent article on the theory of economic growth, Schlüter asks his fellow economists: "Is our theory of growth suitable for application in economic policy making?"[1] His answer is "no" or "perhaps somewhat." Even this eventuality depends upon the occurrence of a very particular condition that is seldom fulfilled: "The introduction of technical advances has become more continuous and the lead time between invention and innovation has become shorter."[2] As we will show, this is in fact not true for basic innovations. The innovative process, "the tiger of technical progress," (Paul A. Samuelson), abhors steady movement. Periods rich in basic innovation alternate with phases of innovative poverty; the process is an alternating trend, and the trend changes during the technological stalemate. This fluctuation is depicted in Figure 4–1 (page 130).

We will now show that inventions and innovations are not closely linked in time. There are innovative surges in which swarms of technical innovations do emerge in close formation. However, in between these surges, there are long dry spells in which there is scarcely any movement in the basic innovative process. Table 1–1 (page 31) indicates that during the twenty years between 1953 and 1973, there have been very few basic innovations or radical technical changes carried out; improvement innovations in established markets have predominated. On the other hand, the years around 1935 produced a bountiful crop of basic innovations; in the 1930s and 1940s, a large number of new industrial fields, of which millions live today, were opened up, including the new synthetics, television, jet airplane, and computer industries. In any event, there can be no illusions about continuity in the process of technical advancement.

The opposite, that is, discontinuity is what occurs in reality. The emergence of basic innovations has not become more regularly paced or more continuous over the past 200 years; in fact, its fluctuations have become more pronounced. The historical stream of basic innovations shows discontinuities, a kind of ebb and flow. These discontinuities will now be illustrated with case material and then the pattern will be tested for reliability.

THE PROCESS OF BASIC INNOVATION

We first need to clarify the term "innovation process." The innovation process is the stream of social and technological innovations that are implemented within a society. In a narrower sense, this process is qualitatively and quantitatively defined by the types and numbers of technological innovations that occur in a society at successive intervals. This concept of the innovation process has been accepted in both theory and practice. The innovation process is called technical progress by neoclassical economists, including the broadening of the available technical knowledge and the application of new knowledge in the form of new products, production materials, and production techniques within the economy.

The term innovation process is preferable to progress for one need not always speak of progress when dealing with technical change and the emergence of new technologies. The process of broadening society's technical knowledge, on the other hand, is called the process of scientific discovery and invention (invention process). This concept involves both the perception of ideas and the generation of new kinds of knowledge. We call the coupling of this process with the innovation process knowledge transfer. It is necessary to distinguish between the invention process, the innovation process, and the knowledge transfer because the time span between basic invention and basic innovation often exceeds the life span of the inventor.

Historically, the industrial development of the West proceeded in the following manner. Applicable scientific discoveries and practicable inventions formed a reservoir of new types of investment opportunities with which the economy eventually eased out of the quandary of successive technological stalemates, which occurred in the decades around 1825, around 1875, and around 1925. Prior to those years, the invention process had provided ideas for new and more efficient technologies in transportation, steelmaking, and so on, that could be implemented to overcome disturbances in the world economy. Today it is the task of the innovation researchers and corporate planners to understand and utilize these processes wisely.

Understandably, people cannot always agree among themselves as to what the proper approach toward innovation should be. They even differ as to what the relevant questions are. May we use new technologies such as nuclear energy, biotechnology, and so on, to provide for new employment opportunities, or must we—for the safety of life and for the conservation of life-styles—restrict innovation in industry more than we do now? What are the risks involved; what are the tradeoffs? Has not the alternative to the promotion of useful industrial innovation always been the stepup of public investment in weapon systems and war industry?

One handicap to innovation research is the difficulty in describing unfamiliar new concepts. The joke circulating among innovation researchers is that we are all like the seven blind men of India who try to describe the sleeping elephant. The blind man standing near the elephant's trunk claims the beast is a kind of snake; the man at its feet says it is more like a tree; and the man touching its ears maintains it is really a cross between an eagle and a bat. The unknown animal generates much vain debate: *tot homines, quot sententiae.*

Another shortcoming of the current discussion about new technologies is that innovation researchers today usually do not distinguish between different types of innovations, and thus there are many pointless misunderstandings. Better definitions will be required to make more reliable assertions about the innovation process in the private and public domains of society. Even the differentiation between basic and improvement innovations used here is only approximate although it serves its purpose.

This chapter will deal only with one aspect of the total spectrum of innovation. At this point, it is appropriate to limit ourselves to the emergence of technological basic innovations rather than attempting to include social innovations. We will therefore also eliminate from discussion the huge class of improvement innovations that rationalize, renovate, or modernize existing production technologies. We are concerned here with the establishment of new forms of technology, not with incidental improvements in the quality of the output or productivity of the inputs occurring in existing technologies. We will also bypass the nontechnical institutional innovations that do not form the scientific hard core of innovative activity. They are, of course, a necessary accompaniment to technological innovations. New technologies often cannot be applied without rethinking organizational change or an amendment to an existing statute. We are therefore proceeding on the assumption that the hard core and the major scientific accomplishments of technical history should be sought in the technological basic innovations.

Fortunately these major scientific events have been well documented so that we have abundant data with which to work. If up to now the economic theory of technical progress has been without empirical relevance, as J. Kornai claims in *Antiequilibrium*,[3] it is not because of a scarcity of data but because *technical change is still terra incognita in modern economics* (Jakob Schmookler).

THE EMERGENCE OF BASIC INNOVATIONS

Technological basic innovations derive from the tree of knowledge in our culture. These basic innovations are the source from which new products and services spring and in turn create new markets and new industrial branches to supply them.

The conversion of scientific-technical knowledge from theory to practice is a process of exploring new areas of unfamiliar activity. This costs time and capital because every experiment, whether or not it is ultimately successful, requires a commitment of these resources. In fact, the researcher's efforts often come to nothing. Of the many ideas conceived from new types of knowledge, very few of them reach technical feasibility; and of the ideas that do, even fewer find an entrepreneur willing to consider the economic realization of innovative projects, much less willing to invest money in the experiments. Finally, only a small percentage of the projects considered and refined are commercially executed. The knowledge transfer from invention to innovation is a selective process in the minds of experts where the stream of inventions is diverted through a series of filters.

This multistage selective process tests the new idea against progressively more exacting feasibility standards and value preferences. It can take many expensive developmental steps, many of which are in the wrong direction, to enable a concept to pass the next more difficult test it must face.

The ideas that pass these tests without being filtered out constitute emerging potential innovations. We will show that there are dramatic fluctuations in the rate of their appearance over time. One must then question what are the sources of this ebb and flow. In this chapter, we will trace the origins of a number of basic innovations, most of which have emerged in surges during times of depression. In Chapter 5, we will attempt to discover the reasons for the abatements between these periods of high creativity. A few important inferences follow from our empirical observations of the innovative process. However, the conclusions drawn from casuistic observations are only as reliable as the methods used to gather the data.

Essentially, we had the same problems as the authors of TRACES[4] who had asked themselves, "how one should select types of innovations which are representative examples for lines of development in history," and "how one should set dates for the starting points and ending points of these historical lineages."

Our team essentially followed the approach taken by the TRACES researchers, that is, answers to these questions require value judgments and presuppose a knowledge of the scientific-technical area involved. The competence to make such judgments can therefore be found in experienced scientists and engineers who are active in these fields. Drs. U. Mende, L. Brache, F. Dorner, C.D. Stolze, and J. Trimborn gathered and classified the scientific information and raw data and chose the critical events from technology history that were to be examined (inventions, improvements in the state of knowledge, and basic innovations). The Fritz Thyssen Stiftung, the National Science Foundation, and the Wissenschaftszentrum Berlin gave financial support for this research.

Thus, our criteria for the selection of basic innovations from the multiplicity of recorded technical events can be expressed as follows: *a technical event is a technological basic innovation when the newly discovered material or newly developed technique is being put into regular production for the first time, or when an organized market for the new product is first created.* For example, the establishment of the first Fuchsin dye factory by F. Bayer in 1860 is the basic innovation for the aniline dye industry, and it was only through Hérault's invention of the cathode oven in 1987 that electrolysis, although a familiar process under laboratory conditions, could be put to practical use in the commercial extraction of aluminum. The history of the telephone also demonstrates how these criteria apply. The basic innovation did not occur with the invention of the Bell apparatus; the first real use of the telephone began in 1881 when the first central exchange was opened in Berlin, allowing connective links to be established within a network of users. The data are presented in Tables 4–1 to 4–7.

Table 4–1. Basic Innovations in the First Half of the Nineteenth Century.

New Concept	*Inno-vation*	*Inven-tion*	*Years Lead Time*	*Tempo of Change*
High voltage generator	1849	1820	29	3.44
Electric impulse inductor	1846	1831	15	6.66
Deep sea cable	1866	1847	19	5.26
Production of electricity	1800	1708	92	1.08
Insulated wiring	1820	1744	76	1.31
Arc lamp	1844	1810	34	2.94
Velocipede	1839	1818	21	4.76
Rolled rail	1835	1773	62	1.61
Pulled wire	1820	1773	47	2.12
Puddling furnace	1824	1783	41	2.43
Coke blast furnace	1796	1713	83	1.20
Crucible steel	1811	1740	71	1.40
First German railroad line	1834			
Locomotive	1824	1769	55	1.81
Telegraph	1833	1793	40	2.50
German customs union	1834			
Lead chamber process	1819	1740	79	1.26
Pharmaceutical production	1827	1771	56	1.78
Quinine production	1820	1790	30	3.33
Vulcanized rubber	1852	1832	20	5.00
Portland cement	1824	1756	68	1.47
Potassium chlorate	1831	1777	54	1.85
Photography	1838	1727	111	0.90

SURGES OF INNOVATIONS

We can now illustrate our discontinuity hypothesis using our data in Tables 4–1 to 4–5, which show what is approximately a representative sample of basic innovations implemented in the Western economies during the past two centuries. We have presented our data as a chronology organized into successive time periods. The time series of the number of basic innovations occurring in the twenty-two ten-year periods between 1740 and 1955 are graphically illustrated in Figure 4–1. The number of basic innovations emerging at a given time clearly fluctuates dramatically. The tide of innovations is often

Table 4–2. Basic Electrotechnical Innovations in the Second Half of the Nineteenth Century.

New Concept	*Innovation*	*Invention*	*Years Lead Time*	*Tempo of Change*
Electrodynamic Measurement	1846	1745	101	0.99
Lead battery	1859	1780	T 79	1.26
Double armature dynamo	1867	1820	47	2.12
Commutator	1869	1833	36	2.77
Cylinder armatured motor	1872	1785	87	1.14
Arc lamp	1873	1802	71	1.40
Incadescent light bulb	1879	1800	79	1.26
Electric locomotive	1879	1841	38	2.63
Electric heating	1882	1859	23	4.34
Cable construction	1882	1820	62	1.61
Telephone	1881	1854	27	3.70
Steam turbine	1884	1842	42	2.38
Water turbine	1880	1824	56	1.78
Transformer	1885	1831	54	1.85
Resistance welding	1886	1841	45	2.22
Arc welding	1898	1849	49	2.04
Induction smelting	1891	1860	31	3.22
Meters	1888	1844	44	2.27
Electric railroad	1895	1879	16	6.25
Long distance telephoning	1910	1893	17	5.88
High tension insulation	1910	1897	13	7.69
Gasoline motor	1886	1860	26	3.84

turning. This ebb and flow in the innovation process is clearly opposed to the theory that innovations flow in a steady stream and thus will appear continuously—and in uniform quantities over time.

The interesting question is how this varying flow in the emergence of basic innovations fits into the long-term economic development. Before we turn to this question, we would like to counter a possible criticism about how representative our data are because the conclusions derived from the data will be no more reliable than the data themselves.

How reliable are our data? We must answer this question in light of how the data in Tables 4–1 to 4–5 are being employed. We are

Table 4–3. Basic Innovations in Chemistry in the Second Half of the Nineteenth Century.

New Concept	*Inno-vation*	*Inven-tion*	*Years Lead Time*	*Tempo of Change*
Thomas steel	1878	1855	23	4.34
Safety matches	1866	1805	61	1.63
Aniline dyes	1860	1771	89	1.12
Cooking fat	1882	1811	71	1.26
Indigo synthesis	1897	1880	17	5.88
Sodium carbonate	1861	1791	70	1.42
Aluminum	1887	1827	60	1.66
Refrigeration	1895	1873	22	4.54
Rayon	1890	1857	33	3.03
Gas heating	1875	1780	95	1.05
Oxyacetylene welding	1892	1862	30	3.33
Dynamite	1867	1844	23	4.34
Chemical fertilizer	1885	1840	45	2.22
Preservatives	1873	1839	44	2.27
Electrolysis	1887	1789	98	1.02
Antitoxin	1894	1877	17	5.88
Chloroform	1884	1831	53	1.88
Iodoform (antiseptic)	1880	1822	58	1.72
Veronal (barbiturate)	1882	1863	19	5.26
Aspirin	1898	1853	45	2.22
Phenazone (synthetic pain-killer)	1883	1828	55	1.81
Baking Powder	1856	1764	92	1.08
Plaster cast	1852	1750	102	0.98
Mass production of sulphuric acid	1875	1819	56	1.78
Synthetic alkaloid (cocaine)	1885	1844	41	2.43
Synthetic alkaloid (chinoline)	1880	1834	46	2.17
High-grade steel	1856	1771	85	1.17

Table 4–4. Modern Technological Basic Innovations in the First Half of the Twentieth Century.

New Concept	*Inven-tion*	*Inno-vation*	*Years Lead Time*	*Tempo of Change*
Automatic drive	1904	1939	35	2.86
Hydraulic clutch	1904	1937	33	3.03
Rollpoint pen	1888	1938	50	2.00
Catalytic cracking of petroleum	1915	1935	20	5.00
Watertight cellophane	1900	1926	26	3.85
Cinerama	1937	1953	16	6.25
Continuous steelcasting	1927	1948	21	4.76
Continuous hot strip rolling	1892	1923	31	3.23
Cotton picker (Campbell)	1920	1942	42	2.38
Cotton picker (Rust)	1924	1941	38	2.63
Wrinkle-free fabrics	1906	1932	26	3.85
Diesel locomotive	1895	1934	39	2.56
Fluorescent lighting	1852	1934	82	1.22
Helicopter	1904	1936	32	3.13
Insulin	1889	1922	33	3.03
Jet engine	1928	1941	13	7.69
Kodachrome	1910	1935	25	4.00
Magnetic taperecording	1898	1937	39	2.56
Plexiglass	1877	1935	58	1.72
Neoprene	1906	1932	26	3.86
Nylon, perlon	1927	1938	11	9.09
Penicillin	1922	1941	19	5.26
Polyethylene	1933	1953	20	5.00
Power steering	1900	1930	30	3.33
Radar	1887	1934	47	2.13
Radio	1887	1922	35	2.86
Rockets	1903	1935	32	3.13
Silicones	1904	1946	42	2.38
Streptomycin	1921	1944	23	4.35
Sulzer loom	1928	1945	17	5.88
Synthetic detergents	1886	1928	42	2.38
Gyro-compass	1827	1909	82	1.22
Synthetic light polarizer	1857	1932	75	1.33

(Table 4–4. continued overleaf)

Table 4–4. continued

New Concept	*Inven-tion*	*Inno-vation*	*Years Lead Time*	*Tempo of Change*
Television	1907	1936	29	3.45
"Terylene" polyester fiber	1941	1955	14	7.14
No-knock gasoline	1912	1935	23	4.35
Titanium	1885	1937	52	1.92
Transistor	1940	1950	10	10.00
Tungsten carbide	1900	1926	26	3.85
Xerography	1934	1950	16	6.25
Zipper	1891	1923	32	3.13

Source: G. Mensch, "Zur Dynamik des Technischen Fortschritts," Zeitschrift für Betriebswirtschaft, 41/1971, pp. 295-314. Data distilled from case studies by Jewkes, J.D. Sawers, and R. Stillerman, *The Sources of Invention*, 1st ed. (London: 1960).

Table 4–5. Basic Innovations (Organizational and Technical) in the Industrialization of Brünn, 1740–1800

New Concept	*Basic Innovation*	*Basic Invention*	*Years Lead Time*	*Tempo of Change*
1. Institutionalization of the mercantilist concept of development policy	1743	1653	90	1.11
2. Establishment of a development bank	1751	1686	65	1.54
3. State investment in textile industry	1762	1701	61	1.63
4. Buildup of viable textile manufacturing plant	1764	1715	49	2.04
5. Forcing labor into the textile industry	1762	1694	68	1.47
6. Technical training of textile workers	1765	1701	64	1.57
7. Restricting guild control over industry	1754	1672	82	1.21
8. Highway connection to Brünn	1740	1704	36	2.72
9. Establishment of central commerce administration	1746	1666	80	1.25
10. Procurement and dissemination of economic information	1754	1666	88	1.13
11. Introduction of a general protective tariff	1775	1737	38	2.63
12. Abolition of the monopolistic privileges for textile factories	1763	1672	91	1.08
13. Departure from rigid economic policy direction	1785	1747	38	2.63
14. Establishment of a market for domestically produced textiles	1769	1727	42	2.38

Source: H. Freudenberger and G. Mensch, "Von der Provinzstadt zur Industrieregion (Brünnstudie); Ein Beitrag zur Politökonomie der Sozialinnovation, dargestellt am Innovationsschub der Industriellen Revolution im Raume Brünn," Göttingen, 1975.

Figure 4–1. Swells of Basic Innovations.

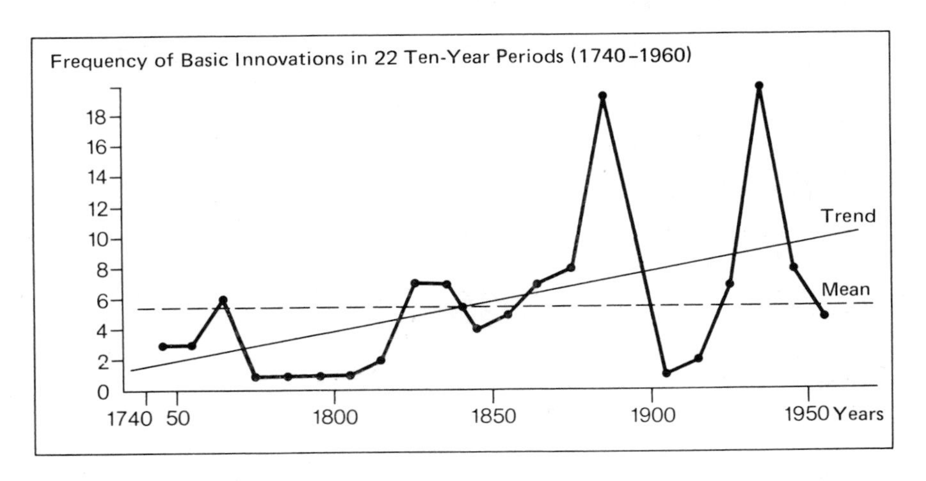

attempting to illustrate our claim that there was only a limited interest in implementing basic innovations during the prosperous phases of the long-term economic process; in contrast to this attitude, the tendency to innovate was very marked during the critical periods of technological stalemate. The exact numbers of innovations are of secondary importance; for our purposes it suffices if we can demonstrate that dramatic changes occur in the quantities of innovations that are produced over time.

Returning to the issue of the reliability of our data, we would still be able to consider our data in Tables 4–1 to 4–5 as representative and continue to work with them even if other researchers who had worked fully independently from us discovered a few more or less basic innovations than we did as long as they noted similar fluctuations, which is indeed the case; see Tables 4–6 and 4–7.

Here we should also point out that we did not discover the phenomenon of surges of basic innovations until after our data had been collected. Originally our research was directed toward a completely different goal from the investigation of discontinuities in the innovative process. Actually, this phenomenon of clustering of basic innovations was a surprise to us. We were also surprised to discover that a comparison of our innovation data with other sources confirmed the information's reliability. The cases collected by other researchers all show that an unusually large number of basic innovations occurred in the mid-1820s, the mid-1880s, and the mid-1930s, whereas the decades in between produced many fewer basic innovations.

The comparison of our data with data from other sources is shown in Tables 4–6 and 4–7. When taken together these findings establish that the innovative process did not open up new industrial frontiers in a continuous fashion; the process was really a surging and receding tide in the economy.

The Discontinuity Hypothesis

We can now pursue the question of how the fluctuations in the frequency of basic innovations fit into the overall economic dynamic that we have demonstrated with the aid of the metamorphosis model (Figure 2–8). The evolutionary alternation between stagnation and innovation allows us to surmise that surges of basic innovations will come during the periods when stagnation is most pressing, that is, in times of depression. The basic innovations put an end to investors' pessimism and their wait and see attitude. The technological stalemate, the pause for adjustment during the economic growth process, is overcome by a surge of innovations that provides new frontiers to enrich the economy. In this way the economy *can* recover from a crisis.

On the basis of the turning events in the 200-year calendar of development of the Western industrial societies, we should view the be-

Table 4–6. The Swell of Basic Innovations in the Period Around 1885—According to Different Sources.

	Frequency of Basic Innovations			
	in 5-Year-Period		*in 10-Year-Period*	
Period	*Source 1*	*Source 2*	*Source 1*	*Source 2*
1865–1869	4	6	4	6
1870–1874	3	5	8	10
1875–1879	5	5		
1880–1884	11	14	19	18
1885–1889	8	4		
1890–1894	4	9	10	12
1895–1899	6	3		
1900–1904	0	1	1	1
1905–1909	1	0		
1910–1914	2	3	2	3

Source 1: Tables 4-2 and 4-3 above.

Source 2: All the significant accomplishments in technical history mentioned by Schumpeter in *Konjunkturzyklen I and II*, Göttingen, 1961.

Table 4–7. The Swell of Basic Innovations in the Period Around 1935—According to Different Sources.

Period	*Frequency of Basic Innovations*		
	Source 1	*Source 2*	*Source 3*
1900-1904	0	18	—
1905-1909	0	16	—
1910-1914	0	29	—
1915-1919	0	34	2
1920-1924	4	29	
1925-1929	3	34	4
1930-1934	7	43	
1935-1939	13	48	8
1940-1944	5	38	
1945-1949	3	23	3
1950-1954	4	20	
1955-1959	1	0	3

Sources: Table 4-4. Mixed data (both basic innovations and improvement innovations have been included) from J. Schmookler, *Invention and Economic Growth* (Cambridge: 1966), pp. 220-222; Schmookler's presentation made it impossible to sift out the basic innovations. All radical innovations requiring a developmental period of fifteen years or more taken from the collection by E. Ulrich and M. Lahner, "Analyse von Entwicklungsphasen technischer Neuerungen," Mitteilungen des Instituts für Arbeitsmarkt und Berufsforschung, Erlangen, February, 1969.

ginning of a period of depression and the beginning of a recovery phase as gateposts through which the surges of basic innovations can pass. If basic innovations really have produced the trend shift from depression to recovery, the data from the Kuznets schema (see Table 2–1) should also indicate in what periods in economic history the swarms of basic innovations appeared. Thus, I conjecture the following periods to exhibit spurts of basic innovations:

1. Pre-1787
2. After 1814 and before 1828
3. After 1870 and before 1886
4. After 1925 and before 1939
5. After 1983 and before 1995[a]

[a] See Chapter 7 for prediction of the next surge of basic innovations, which is given by the frequency distribution (μ, σ) where μ = mean year (1989) and σ = standard deviation ($\pm$ 5 years).

Do the innovation peaks in Figure 4–1 fall within the intervals we have just set out? Yes, they do. The discontinuity hypothesis therefore makes sense; most of the basic innovations occurred when the socioeconomic climate (during a technological stalemate) would have led one to expect them. The appearance of significant numbers of innovations in the periods we have named is therefore very plausible.

However, there is a clear distinction between plausible occurrences and scientifically proven facts; thus we should be forewarned to proceed with caution. The discontinuity hypothesis has wide-ranging consequences; we can do justice to them, however, only if the discontinuity hypothesis itself can withstand critical attempts to disprove it. We must therefore go beyond statements of plain plausibility and test the validity of our hypothesis.

Do Basic Innovations Occur Randomly?

A significant argument against the reliability of the discontinuity hypothesis must be countered before we can trust it. An argument against any regularity in the reappearance of innovative surges could be made simply by denying that there is any adequate explanation for the phenomenon. Such a counterargument has often been applied in the economic literature, especially in the literature on evolutionary processes and long-term cycles. It is the same argument that Weinstock raised against the model of Kondratieff cycles. Weinstock's reservations about the concept of long waves stem from the alleged lack of a theoretical foundation: "there has been no plausible explanation found for the uniform, inevitable return of the long waves."[5] However, my metamorphosis model provides the theory of basic innovations coming in clusters (but not necessarily in the rhythm of a uniform wave). Therefore, lack of explanation is no counterargument against the discontinuity hypothesis.

But that was only a weak defense of our hypothesis. The strongest defense is disproving the "null hypothesis." Long ago, David Hume taught that "nothing follows from following" unless there is a theory that explains the appearance of all of the elements in a sequence. Hume's logic then implies that the lack of a theory is a warning not to draw any wide-ranging conclusions from the consequences of an event, for if there is no valid theory explaining the cyclical recurrences, it simply means that one can and must view the ebb and flow of innovations as the work of chance. This exactly is the null hypothesis which we test and disclaim in the following.

On the other hand, because as Karl Popper has shown one can never actually prove a theory, a shadow of doubt would still remain even if there were a theoretical explanation for the recurrence of innovative surges. Therefore, with or without a theory, we must accept the fact that a trace of doubt is always justified if a hypothesis provides a plausible explanation for certain observed facts of life. To be sure, one would like to know the degree of the doubt in order to estimate how secure or insecure one will be in trusting one's hypothesis. An evaluation of the risks of error that one should take into account when working with theories is therefore necessary to insure that one is on a promising line of investigation. In the realm of science, "a pinch of probably is worth a pound of perhaps" (James Thurber, *Lanterns and Lances*).

The counterargument to the discontinuity hypothesis states that observed surges of innovation are due to nothing more than chance. One can test this null hypothesis with the help of nonparametric statistics, that is, with the runs test. This is what we have done with our data on basic innovations.[6]

Thus, we have calculated how often one must apply a chance mechanism before it will produce an unusual pattern such as the pattern that occurs in the variations in the innovation stream. By analogy, when the change mechanism produces such a pattern only twice or three times in 100 applications, the risk of error is then about 2 or 3 percent when one relies upon the discontinuity hypothesis. So that no error slips past us because the discontinuity pattern depends upon the chosen time divisions, we have also tested the time series that arise when we assign the data into one-, two-, three-, and other year periods, respectively. The separation of the data into ten-year periods produces the discontinuity pattern shown in Figure 4–1. Table 4–8 gives the probability that the particular pattern in the time series is due to chance (risk of error).

We can now judge the counterassertion to the discontinuity hypothesis on the basis of these probabilities. Irrespective of the periodization scheme, the likelihood that the observed discontinuities are simply the work of a random process is well under 5 percent, fluctuating mostly around 2.5 percent. Thus we can reject the null hypothesis with a risk of error of approximately 2.5 percent, which means that the discontinuity hypothesis has stood up against the test of chance with great significance.

The consistent pattern of ups and downs in the innovative stream must therefore be treated as an assured empirical fact. The level of significance (around 97.5 percent) is so high that the discontinuity pattern would not loose its form even if some innovations are being taken from the list of data (Tables 4–1 to 4–5), for example, in response to debate over certain data points. Thus, the empirical result is robust.

Table 4–8. Test for Randomness in the Discontinuity Pattern in the Time Series of Basic Innovation between 1740 and 1960—Results.

Different Periodization Schemes		*Probability That the Discontinuity Pattern in the Time Series Can Be the Work of a Random Process*
Length of the Time Periods	*Number of Periods*	
1 year	220	0.0179
2 years	110	0.0351
3 years	73	0.0091
4 years	55	0.0294
5 years	44	0.0344
6 years	36	0.025 (–)
7 years	31	0.025 (+)
8 years	27	0.025 (–)
9 years	24	0.025 (–)
10 years	22	0.025 (+)

Time Series with more than 40 periods are "long series"; their probabilities can be calculated precisely. There are tables used to calculate for "shorter series," from them one can determine whether the "level of significance" is less than 0.025 (–) or is greater than 0.025 (+) but less than 0.05.

Source: S. Siegel, *Nonparametric Statistics for the Behavioral Sciences* (New York: 1956), p. 252.

SUMMARY

The popular opinion that technical progress takes place during modern times in a continuous fashion (the continuity hypothesis) is inconsistent with reality. According to this widely held view, a steady stream of innovations should be holding the forces of stagnation in check within our industrial societies. This continuity hypothesis has already been contradicted by the crisis points occurring during past years alone. In contrast, the discontinuity hypothesis postulates a dramatic alternation between periods of innovative abundance and innovative scarcity. The changing tides, the ebb and flow of the stream of basic innovation explain economic change, that is, the difference in drive in growth and stagnation periods.

The discontinuity hypothesis has been verified using the statistical method. It stood up well in the tests and has been shown to be relevant for long periods in the past. The lack of basic innovations seems to have caused the slide into past technological stalemates, and innovative surges finally brought the stalemate to an end. The last time when this occurred was during the 1930s when the great depression was overcome.

In Chapter 7 I give a rough estimate of the next swarm of basic innovations which I am fairly convinced of coming in the forseeable future.

REFERENCES

1. K.P. Schlüter, *Zeitschrift für die gesamte Staatswissenschaft*, 129 (1973), pp. 613–633.
2. Ibid., p. 614.
3. J. Kornai, *Antiequilibrium* (Amsterdam: 1971).
4. Illinois Institute of Technology Research Institute, "Technology in Retrospect and Critical Events in Science," *NSF*, 1, p. xi.
5. U. Weinstock, *Das Problem der Kondratieff-Zyklen* (Berlin-Munich: 1964), p. 120.
6. G. Mensch and C.D. Stolze, *Innovation und Industrielle Evolution, Preprint I/73–29* (Berlin: International Institute for Management, 1973).

Chapter 5

The Shortage of Basic Innovations: Due to a Lack of Scientific Creativity?

"It is no one's fault, but everyone's problem."

(Robert F. Wagner, Jr., commenting in July 1965 on the acrimonious debate over the water shortage that threatened New York City during the final months of his last term as mayor).

Technological basic innovations are events of putting scientific results into economic practice. If there are too few of these events, the industrial economy will suffer from stagnation. Why are there so few basic innovations in times of prosperity? Did investors shy away from basic innovations? Or did science fail to provide useful concepts? Let us put these questions in this way: Can the economic leadership shift the responsibility for neglecting basic innovations in good times onto the shoulders of science? In particular, can the economic leadership excuse itself for not investing enough in basic innovations by claiming that the scientists did not produce enough innovative ideas? Is there some justification for the complaint raised by the top manager of one research-intensive company gripped by stagnation—"We are not getting any workable ideas?" Is there really a lack of practically relevant theories and the deficiency of new ideas during times of stagnation?

In this chapter we will examine the question of who should be held responsible for the lack of basic innovations. We will investigate the possible contention raised by the economic leadership that this lack was not the fault of leadership, but rather it was caused by an intermittent lack of creativity, and thus gaps existed in the economy's continuing supply of workable innovations. If this view were justified, blame for the development of stalemates in technology and the tendency to fall into crises would ultimately devolve upon researchers, scientists, and the institutional network of their community.

In the preceding chapter for the 220 years surveyed, the empirical findings showed that surges of technological basic innovations emerged after economies had fallen into a serious crisis and then passed through years of depression. A graphic representation of the fluctuations in innovative implementation shows that there is a damming up of innovative activity until the onset of economic crises, and then innovations break through the floodgates. This picture leads to only one conclusion. Since the Industrial Revolution whenever the economy changed from high growth to low growth in leading sectors, the market mechanism has not been able to react quickly enough in drawing capital and labor out of the stagnating sectors and directing them into new lines of real investment where new demand and new technologies offer new kinds of occupations. In any event, the redirection of resources into new types of occupation did not suffice to offset the stagnating trends in many traditional areas.

Instead, we saw that every fifty years when stagnation developed, the economy first reacted adversely by cutting down research and development and experimentation. Instead of redirecting it, firms released large amounts of capital and manpower. Instead of increasing expenditure in research and development, firms saved on investment in innovation. Only after the crisis hit the capital market did investors wake up. In the past, the swarm of basic innovations always appeared simultaneously with depressions but following a world economic crisis. In the past, the capital market mechanism brought an end to the technological stalemate only after having been shocked, and the market economy provided new and different kinds of employment for the jobless only after the government stepped in.

Technological and nontechnical basic innovations are indispensable prerequisites for a leveling out and shift toward an economic upswing. An upswing above all requires new types of information, given that labor and capital are not the minimum factors. In times of technological stalemate, useful technological information is a scarce good—not because it is lacking but because the market for such information does not exist or does not function. The information market is imperfect and disorderly. Order requires organization, and organization requires an energy supply. Forward movement of the economy requires work at the bottleneck. This is the secret behind the so-called self-healing powers of the economy. Without investment in innovations, the forces of stagnation reduce growth to zero or even cause a negative growth rate. Without an orderly supply of concepts for basic innovations, the economy becomes bogged down in economic crises. The trouble is that capital market mechanisms are not always a reliable substitute for the nonexisting information market mechanism.

I maintain that the major cause of the delayed introduction of basic innovations is not lack but immobility of precious information. What is lacking is the information-processing capacity for improving the transfer of precious knowledge. The information market needs its own market mechanism. If the crisis-prone character of the market economy is caused by insufficient market clearing in the information market, there are certainly corrective aids available.

Placing the blame on the capital market mechanism is neither a new nor even a leftist idea. Karl Marx was not the first to trace the crisis tendency in the profit-oriented economy to a built-in conflict between the established, but stagnant, lines of production, and the repressed, but highly promising, productive forces; in short, to losses from friction within the system. The liberal literature also does not overlook the imperfections in the market economy. In his famous essay on economic welfare and the allocation of resources for invention, Kenneth Arrow states his beliefs that

- a free-enterprise economy will underinvest in invention and research (as compared with an ideal) because it is risky, because the product can be appropriated only to a limited extent. This under-investment will be greater for basic research than for applied research;
- the incentive to invent is less under monopolistic than under competitive conditions but even in the latter case it will be less than is socially desirable; and there is a bias against major inventions.[1]

Underinvestment in research, development, and innovations is not the only problem, however. Nearly 200 years ago, Jeremy Bentham, the Utilitarian, pointed out that the profit motive encourages the wrong kind of innovative investment, which can then lead to serious crisis. In his famous letter to Adam Smith "on the discouragement of progress in the inventive industry" (March, 1787), Bentham argues against the market's "tendency to pick out the good projects from the bad, and favor the former at the expense of the latter." Since then, economists have been aware that the profit motive as an incentive to innovate sometimes produces short-term, shortsighted measures that are possibly inconsistent with long-term needs, while long-range strategies and farsighted measures are often neglected by firms and may be completely absent from the economy. Even right-wing critics deplore the confusion of caution with precaution that is so often revealed in capital market decisions.

This contradiction in the market economy system is not a secret. However, this systemic weakness requires a cure, both because the weakness is linked to the recurrent danger of crisis and because it would be foolish to watch a system break down if it can be repaired.

Is it really necessary for a stagnating economy to go through a structural crisis before it can produce basic innovations in any quantity? No. It is conceivable that new forms of activity could enter the economic mainstream without the periodic breakdown in many of the old activities. This would mean, as the Pigou effect implies, that the crises we have observed were tragic and unnecessary exaggerations of the necessary adjustments demanded by the situation.

Laxity is all that is necessary to cause an economy to slide into crisis. Benign neglect permits disorder to spread, disorder is a lack of organization, and disorganization sets energy free. The breakdown nourishes itself; it does not require any outside help. This result follows from the economic entropy law. Benign neglect of basic innovations suffices to shift the trend toward disorder and crisis. Therefore, it is very likely that the major economic crises of the past were due to negligence by the economic leadership in reacting too slowly to counter the forces of stagnation with basic innovations.

If one accuses a certain group of negligence, that group will attempt to shift the blame to another group. It is indeed conceivable that the source of all the economic problems lays in the fact that there were not enough innovative possibilities available at the critical moment because researchers had not been producing enough innovation projects at the appropriate time (*echo effect*). This possibility will now be explored. We would like to know if the lean years for basic innovations are simply an echo of previous lean years in scientific progress. This requires a direct correlation of innovation with invention, which can be expressed as a linear model of the knowledge transfer.

IS THERE AN ECHO EFFECT IN THE TRANSFER OF KNOWLEDGE?

We will now investigate the hypothesis of the echo effect. On the one hand, do ebbs actually occur in the advancement of scientific knowledge that later, possibly even decades later, lead to acute shortages of innovative projects during technological stalemates? And on the other hand, do increases in scientific creativity in later periods produce a high volume of innovations in their wake?

This model of knowledge transfer is linear. It is based on the premise that the movements occurring in the stream of knowledge are mirrored in the stream of innovations after a relatively constant lead time. This simplicity is not naivete; for if it conforms to the facts, the simplest model allows the most powerful conclusions to be drawn.

If the echo effect in the transfer of knowledge could be demonstrated, the economic leaders could no longer be reproached for having failed to take adequate measures to prevent a crisis from developing. Moreover, a theory that pervades the sociological and political literature on economic and social changes would also be proved groundless. The theory charges that prevailing power elites suppress the innovative proposals detrimental to their interests in order to secure the status quo against the inroads of change and that this is the cause of ensuing imbalances in social and economic forces.[2] This imbalance builds up and finally erupts as a crisis. Just like an earthquake that occurs to relieve the tectonic pressure; the force of the societal or economic eruption is proportional to the degree of congestion and inversely proportional to the previous speed of adaptation.

Thus, if the echo effect could be verified on the basis of the historical data on basic inventions and innovations given in Chapter 4, the blame for having caused the crisis would fall on the scientific apparatus, that is, upon the scientists and "the deans of the invisible colleges."

By using a linear model of knowledge transfer and the above data and by either proving or disproving the existence of the echo effect, we can determine which of the elite groups can *not* be saddled with blame for the economic crises.

SURGES OF SCIENTIFIC ADVANCEMENT

Is there an echo effect in the transfer of knowledge? Can the slack periods and surges in the innovation process simply be traced to prior slack periods and surges in scientific advancement? The fundamental question that must be answered first, however, is whether or not there really are surges of knowledge.

In order to deal with these questions operationally it is helpful to make use of the dualistic concept of body and soul. The transfer of knowledge can be understood as the transition from theory to practice within this two-sided world. Acquiring scientific knowledge would then be the invention process within the realm of science and basic research, and the innovation process would be the process of applying the theoretical knowledge within the economy. "Man's manufactures are products of his culture; they are learned skills; they are knowledge."[3]

If we wish to test the echo effect for validity, we must describe the scientific learning process in quantitative terms. That means counting or evaluating certain societal accomplishments. If we need to measure the volume of scientific insights emerging in a given time period by the counting method, once again it is sensible to limit that what counts to actual scientific achievements as they are most completely documented. This limitation does not mean we are ignoring the humanities and the arts or, for example, the sociological insights. However, their fluctuations in volume seem to move parallel to the volume of accomplishments in the natural sciences because both types of insights flourish according to the strength of the cultural elan of each period. By the same token, we will also not deal directly with events occurring on the so-called archaeological plane of knowledge, which Michael Foucault defines as "a plane beyond the scientist's consciousness which forms, however, an equally important aspect of his scientific conduct."[4] The activities occurring on all cultural levels during any period are intertwined, and as the scientific learning process is channeled from the archaeological to the epistemological plane, it finds expression in certain basic assumptions and in a certain scientific terminology. "The arts and inventions of each period are only its costumes," Emerson wrote in his essay on self-reliance.

Thus even if we limit our examination to scientific insights, we will also indirectly encompass the insights gained outside of the scientific realm that produce feedback on the scientific learning process. Scientific and cultural insights are closely interrelated: "History is continually shaped by creative spirits; by those who have behind them a different world from the one in which they first found themselves," wrote Heinz Ohff in his tribute to Picasso. Just as technical basic innovations are accompanied by a multitude of nontechnical changes in modus operandi, scientific insights are also accompanied by a multitude of general learning and unlearning experiences. Our problem is to find a countable subset that is approximately representative of the whole set of events. This subset is our list of basic inventions.

We return now to our questions whether or not there are barren and fertile periods experienced in the epistemological fields of scientific learning and if there are perceptible surges in insight. For historians of science, the question touches upon a highly debated issue that has occupied historiographers of science for many years. The uniformists, who believe in a steady cumulative expansion of knowledge, tend to respond negatively while the catastrophists answer affirmatively.[5] For the catastrophists, scholarly opinions that previously were unquestioned lose their credibility when they come into conflict with actual fact. In this situation if a scientist then develops a better theory, catastrophists feel that the entire body of knowledge must be reexamined; certain essential components must be abandoned and others must be radically revised.[6] Usually, scientists view the need for discarding old theories and the burden of having to revise their textbooks as revolutions in their disciplines,[7] a revolution that may easily produce a swell of basic inventions in its wake.

How can one visualize these discontinuities as barren or fertile periods for some epistemological fields? Concreteness is a prerequisite for any new type of empirical work, including this venture into the unresearched area of knowledge transfer. Successive generations of trees of knowledge anchor their roots in epistemological fields. Distinctive variations occur every generation; the spirit of each epoch (zeitgeist) gives to the basic inventions of the epoch an unmistakable character. This distinct quality can be perceived even in the first inventions of the new epoch. That is a vital point—the new kinds of inventions set new quality standards in theory (explanatory power) and practice (technical performance).

Thus, the first fruits of the new harvest from the tree of knowledge arouse a commotion in scholaraly circles and unleash stormy debates over their new character. Some scientists recognize the power of the new principles at first glance while others see no more than Galileo's colleagues did when looking through his telescope, which was nothing. The controversial new theory is a candidate for a new paradigm. The intensity of the controversy is proportional to the scientists' dissatisfaction with the gap between the known facts and the prevailing theoretical structure. When theory and reality deviate too widely, any reasonable new theory creates considerable unrest in the scientific community. Thomas Kuhn has explained this process of scientific revolution as a substitution of the prevailing paradigm by another.[8]

A paradigm is a particularly powerful theory. Kuhn characterizes the main function of paradigm as follows. A paradigm induces the scientists to consider only a fraction of the unending stream of problems as plausible subject matter for their research, that is, it sets priorities. Moreover, the problems that scientists decide to focus on once they accept a particular theory provide new challenges for further scientific insights. New problems become visible, and new, surprising scientific solutions spring from the new paradigm. These new scientific insights are the products of a given epistemological field; they provide valuable information and basic inventions that can be transferred from theory to practice. Therefore, we should be able to observe scientific revolutions that eventually lead to a spurt of basic innovations in practice by observing data on basic inventions as listed in Tables 4–1 to 4–5.

To delineate the concept of a basic invention more precisely, let us use Schmookler's distinction between a subinvention and an invention, first clarifying what these terms mean in the practical world. A subinvention is an improvement in a product or production process that causes little surprise among people in the field because they could have easily discovered it themselves. An invention, on the other hand, is a formula for a new type of product or process that astounds even the experts in a field.[9] Beyond this, a basic invention is more deeply rooted in the world of scientific discovery; it contains a strong general theoretical element. A basic invention usually is the foundation for several inventions and many subinventions.

One can now easily conceive that the rate at which scientific insights develop within a particular field is in no way constant. On the contrary, like every evolutionary process, it is subject to the dynamics of creation, growth, and decline. This organic process of unfolding and maturing thus creates surges of basic inventions. This process starts slowly because the instant that a new theory appears claiming to provide a better explanation for pertinent facts, scientists cannot yet see the full range of problems presented by the theory. The new problem areas will be recognized gradually, and the swelling number of problems will thus be solved gradually. When a set of circumstances is perceived as a solvable problem or as the basis for a solution to a principal problem, it is called a basic invention.

As long as the basic inventions result from knowledge that is still young, they appear in small numbers. Over the years, they increase in number because more scientists see the new paradigm as correct and therefore accept it as a guideline in choosing problems for research. They are rewarded with success because they are working on the basis of a better paradigm than others are. Thus the process is self-reinforcing, ultimately producing surges of basic inventions.

As the masses of originally and subsequently perceived problems within a given theoretical framework are solved, the numbers of new insights emerging in this area gradually diminish. They can stop altogether if this problem area has been exhaustively investigated or if scientific curiosity has been lured in another direction and is now following the parameters of another paradigm, leaving the earlier epistemological field barren.

Over the years, the number of basic inventions vary according to whether the epistemological fields at the frontiers of knowledge remain fertile and productive or—as is today the case in pharmaceuticals—become depleted. It is therefore conceivable that scientists are periodically unable to find practical solutions to important practical problems. Therefore, surges of basic inventions could alternate with periods in which only few basic inventions come out of a particular epistemological field. In fact, there are excellent contemporary examples of a temporary exhaustion of epistemological fertility. The crisis in modern physics is a good example. Even some administrators have recognized that many of the current research projects are unpromising.

The past provides excellent examples to illustrate the dynamic aspects of the invention process. We will use the fields of chemistry and electrophysics as examples to trace the transfer of knowledge from the scientific revolution to the clustering effect in the industrial application of the insights derived from it. The process begins with the introduction of a new theory that quickly becomes a paradigm for further research and produces a swell of investigations and findings. Let us quote T. Kuhn:

> What Lavoisier announced in his papers from 1977 on was not so much the discovery of oxygen as the oxygen theory of combustion. That theory was the keystone for a reformulation of chemistry so vast that it is usually called the chemical revolution.[10]
>
> The discovery of the Leyden jar displays all these features as well as the others we have observed before. When it began, there was no single paradigm for electrical research. Instead, a number of theories, all derived from relatively accessible phenomena, were in competition. None of them succeeded in ordering the whole variety of electrical phenomena very well. That failure is the source of several of the anomalies that provide background for the discovery of the Leyden jar.[11]
>
> [Both] our most recent examples show that paradigms provide scientists not only with a map but also with some of the directions essential for map-making. In learning a paradigm the scientist acquires theory, methods, and standards together usually in an inextricable mixture. Therefore, when paradigms change, there are usually significant shifts in the the criteria determining the legitimacy both of problems and of proposed solutions.[12]

Both examples illustrate record yields in subsequent scientific discoveries. Following the chemical and electrophysical revolutions and the breakthrough in the formulation of a truly workable theory, chemists and electrophysicists could continually develop new problem definitions and suggested solutions that (basic inventions) are the direct results of the scientific revolutions referred to above.

We can now continue with our description of the transfer of knowledge in the fields of chemistry and electrophysics. The tables in Chapter 4 list the dates for basic inventions and basic innovations. These dates pinpoint the beginnings and the end products, respectively, of the transfer of knowledge. They signify pairs of related events; the earlier event is part of the stream of basic inventions and the more recent event merges into the stream of basic innovations.

The basic invention processes for electrophysics and chemistry are represented in Figure 5–1. They are depicted as a time series showing the number of basic inventions per decade according to the listings in Tables 4–2 and 4–3.

The time series in Figure 5–1 show:

1. Strong surges of basic innovations in the electrotechnical and chemical industries; compared to the surrounding decades, the 1880s show a clearly accentuated high level of innovative activity.
2. The frequency distributions of the basic inventions leading to basic innovations in electrophysics and chemistry extend over a very long time; up to 100 years have passed between the first formulation of a paradigmatic theory and the commercial application of the knowledge in industrial practice.
3. These frequency distributions show some fluctuations in the emergence of scientific basic inventions in the areas of electrophys-

Figure 5–1. Frequency of Basic Inventions and Innovations in Chemistry and Electrotechnique Before 1900.

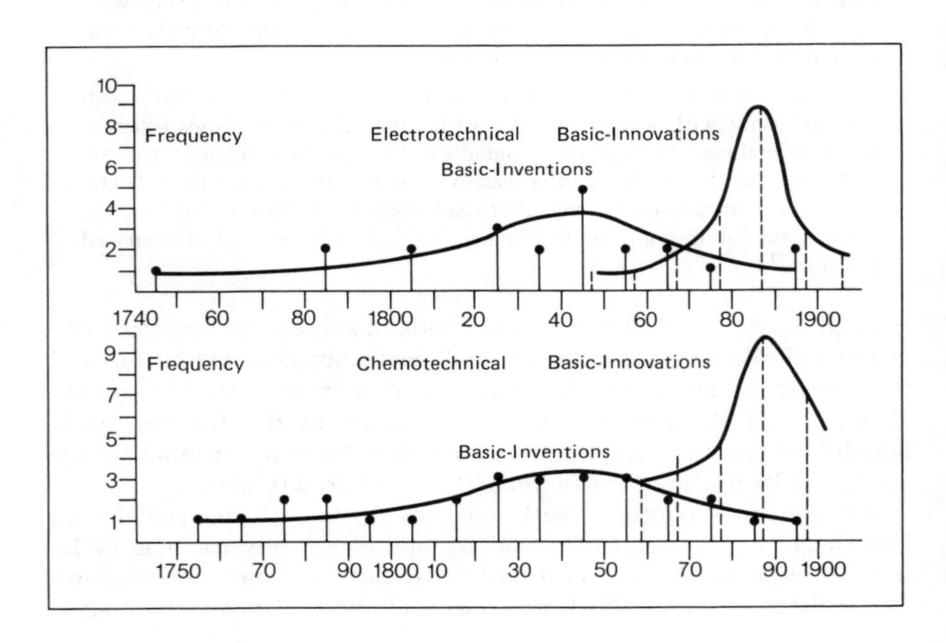

ics and chemistry; with a little imagination one can see a high point in the number of basic inventions around 1840, but with a little skepticism one can easily dispute this creative surge.

Figure 5–1 gives the impression that slack periods and surges in the invention process in fact do exist, but they do not appear to alternate with the same distinctiveness as the discontinuities in the innovation process do. This impression is strengthened as we examine more data.

Let us first examine the basic inventions that provided the basis for the early industrial basic innovations clustering in the years around 1825. The appropriate chronologies from Table 4–1 are represented in Figure 5–2 (bottom). From these data in Figure 5–2 one derives a stronger sense of ups and downs in the stream of basic inventions than one did from Figure 5–1. However, it is questionable whether one really can speak of distinct slack periods and surges in basic inventions.

Now let us examine the basic inventions that provided the scientific foundation for basic innovations in the first hald of the twentieth century. The appropriate chronologies from Table 4–4 are rep-

Figure 5–2. Basic Inventions and Basic Innovations *(First Half of Nineteenth Century).*

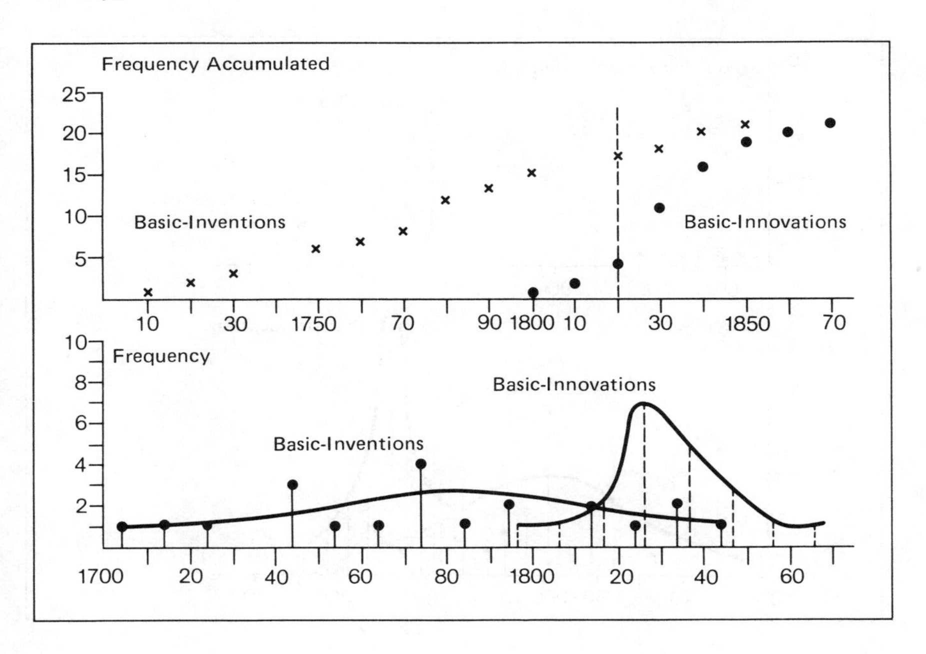

presented in Figure 5–3 (bottom). These frequency distributions strengthen the impression that powerful surges occurred in the stream of basic inventions that may later have led to the surge of basic innovations in the years around 1935. However, it is not easy to judge if slack periods in the stream of basic inventions actually occurred and were later echoed in the scarcity of basic innovations during the years of technological stalemate before the world economic crisis of 1929. Because of the varying optical impressions, simple visual inspection of the above graphs does not allow one to conclude unilaterally either that an echo effect actually exists or that it is not in evidence here. This requires making a methodical survey of the stream of basic inventions.

CHECKING FOR ECHO EFFECTS

Given that the existence of ebbs and flows in the stream of basic inventions cannot be disputed and thus the economic leadership may have an excuse for the occasional scarcities in basic innovations, we must now examine the question of whether an echo effect exists in the process of knowledge transfer. The question must be answered

Figure 5–3. Basic Inventions and Basic Innovations *(First Half of Twentieth Century).*

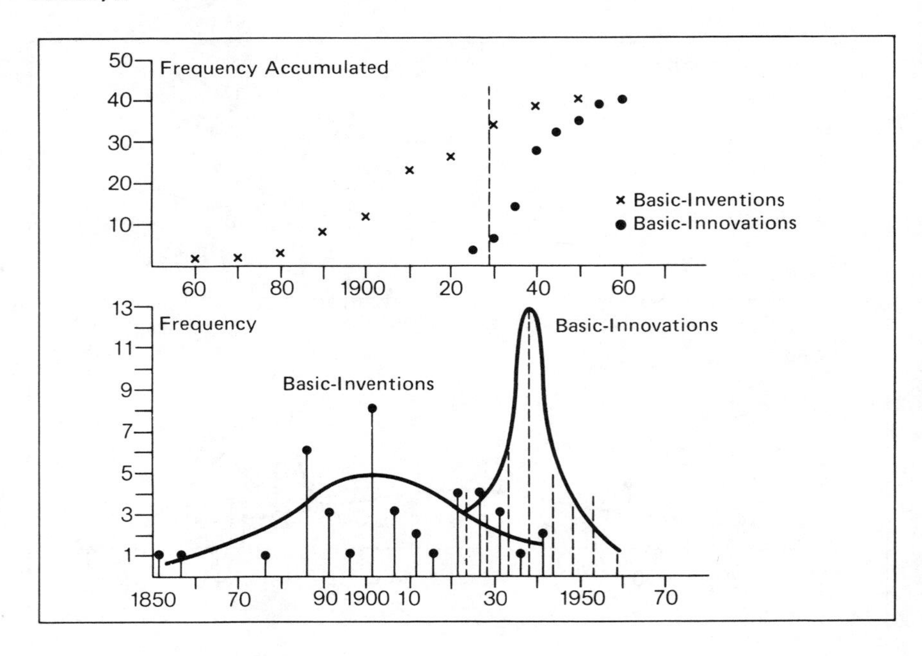

negatively if we can establish no similarity between the stream of basic inventions and the stream of basic innovations. "Similarity" means that the time series for basic inventions as a whole must reveal a pattern of fluctuation that, having taken the temporal displacement into account, corresponds to the discontinuity pattern in the innovative process both in the height and depth of the peaks and troughs and in their distance from each other. Otherwise, the existence of an echo effect could not be substantiated. The time series of all basic inventions would have to exhibit ups and downs similar to those in the emergence of basic innovations for one to be justified in shifting the blame for the scarcity of basic innovations onto science and research.

On the other hand, it is unlikely that slack periods in the stream of basic innovations would have developed from corresponding earlier slack periods in the stream of basic inventions if changes in the stream of basic inventions do not exhibit a distinct pattern in the way in which changes in the stream of innovations do. Using the same method that we used to test the discontinuity pattern in the time series of basic innovations in Chapter 4, we can evaluate the discontinuity in the stream of inventions by testing for randomness. We will estimate the likelihood that the pattern of changes over time in the stream of basic inventions is due to chance. Chance alone would be very unlikely to produce sharply defined, regularly recurrent discontinuities. Therefore, a blurred irregular discontinuity pattern in the stream of basic inventions can*not* be considered similar to the significant pattern of discontinuity in the stream of basic innovations if the likelihood of chance producing the former is far above 5 percent for the likelihood of chance producing the latter pattern is significantly below 5 percent.

If we compile all of the invention data in Tables 4–1 to 4–5 into a long time series, the probability of chance being responsible for the irregularities varies from as high as 15 percent (when we use ten-year periods) up to 33 percent (when using five-year periods). Thus, the pattern could very easily have been produced by chance. Clearly, one cannot consider the time series of basic inventions to have a significant similarity to the time series of basic innovations; at best the similarity is superficial.

The concept of an echo effect would allow the scarcity of basic innovations during years of stagnation to be explained by an earlier lack of basic inventions, which in turn was caused by scientists' and researchers' inadequate creative powers. On superficial examination of the data this might appear to be a plausible explanation of the circumstances, but when one methodically examines the facts, this impression is not significant enough to serve as an alibi for the economic leadership.

In fact, whenever technological stalemates have occurred in the past they have not been caused by a lack of new and useful theoretical knowledge but to the economy's apparent inability to translate this knowledge into practical terms. This is the paradox of unused technologies that we observed before 1825, before 1873, and before 1929 and which, according to Marshall Robbins's and others' studies on federal incentives for innovation, can also be observed today.[13]

Today, as has occurred at several points throughout history, numerous useful concepts for new technologies have been perfected in research sectors. However, these technologies are not being applied, at least for the present, within the stagnating economy. This is the result of inertia in factor reallocation.

SUMMARY

One possible excuse offered by the economic leadership accused of responsibility for the lack of basic innovations during the technological stalemate is that the scientists were uncreative or were not sufficiently creative at the right time and did not supply enough innovative ideas when they were needed (echo effect between lack of basic invention and lack of opportunities for investment in basic innovation (theory of lacking technological information).

According to this theory, if it were valid, the blame for the failure to adjust for stagnations should be shifted from the market process to the science and research. Based on the empirical evidence, the echo effect must be rejected. The economic leadership has not been exculpated from the lack of basic innovation. The causes for the lack of innovation are not the lack of useful information but the malperformance of the information market and inertia in capital reallocation. "Calm times teach few lessons. Adversity is a great teacher" (Otto Eckstein[14]).

REFERENCES

1. Kenneth Arrow, "Economic Welfare and the Allocation of Resources for Invention," in R.R. Nelson (Hrsg.), *The Rate and Direction of Inventive Activity* (Princeton: Princeton University Press, 1962), pp. 600–625.

2. E.M. Rogers and F.F. Shoemaker, "Communications of Innovations," New York 1971, p. 341.

3. W. Goldschmidt, *Exploring the Ways of Mankind* (New York: 1960), p. 115.

4. M. Foucault, *The Order of Things* (London: 1970), SXI.

5. W.I. Cannon, "The Uniformitarian—Catastrophist Debate," *Isis*, 51 (1960), pp. 38–55.

6. J. Agassi, Towards a Historiography of Science (Den Haag: 1963).

7. G. Buchdahl, "A Revolution in Historiography of Science," *History of Science* (1965), pp. 55–69.

8. T.S. Kuhn, *The Structure of Scientific Revolutions* (Chicago, 1970).

9. J. Schmookler, *Invention and Economic Growth* (Cambridge: 1966, 1967).

10. T.S. Kuhn, op. cit., p. 56.

11. Ibid., p. 61.

12. Ibid., p. 109.

13. M.D. Robbins, *Federal Incentives for Innovation* (Denver: Denver Research Institute Report C790 to the National Science Foundation, 1973), p. 4.

14. Otto Eckstein, *The Great Recession* (1973–75), Amsterdam 1978, p. 1 and 139.

 *Chapter 6*

Neglect and Hastiness

"Time is
Too Slow for those who Wait,
Too Swift for those who Fear"
(Henry van Dyke)

We have pointed out the consequences of a negligent attitude toward the transfer of knowledge into basic innovations during an economy's growth phase. This means that *today's need for new investment opportunities cannot be satisfied immediately with innovative projects that are both practical and sizable enough* to compensate for stagnation and negative growth in many traditional branches of industry.

Why are the innovative potentials not being used in time? In part, the delay results from the overly categorical division of labor between the research institutions on one end and the places of production on the other. As they pursue different objectives, the separation obstructs the flow of precious information on future needs and on new ways to serve these needs profitably. Innovative possibilities are easily buried by this system; they are only unearthed when stagnation calls the traditional modes of production into serious question. The market mechanism functions well and smoothly in supplying the population with the standardized goods for daily use. Its weakness lies in its inherent difficulties in drawing forth innovative possibilities from the research sector and then directing them to the areas where the need for basic innovation is the greatest. That means, the information market does not function as efficiently as is necessary for industrial economies, which depend on the elastic supply of precious information.

In the past, it has mainly been during the straits of an economic crisis that the economic leadership finally, and after years of delay, has succeeded in organizing the distribution of basic innovation possibilities among those actually able to innovate them. During these emergencies, hastened by the pressure of the times and encouraged by a strong-minded government head (Metternick, Bismarck, or Roosevelt, for example), innovative surges finally occurred in the years around 1825, around 1886, and around 1935.

Based on the paradox of unused technologies, we will now show the economic leadership's neglect of basic innovations in times of prosperity and the subsequent slide into the technological stalemate. We will also demonstrate the problems engendered by the leadership's hastiness as it begins to implement measures to pull out of the technological stalemate.

THE PARADOX OF UNEXPLOITED TECHNOLOGY

In the stagnation phase of the economic process, the lack of basic innovations turns the technological stalemate into a critical situation. Today the economy clearly needs both technical and organizational innovations, not to mention social innocations that are required in the society at large. M. Robbins writes on this paradox of unexploited technologies:

> In view of this apparent need for an increased rate of technological innovation, it is paradoxical that both in the U.S. and elsewhere there appears to be no dearth of technology for increasing the rate of innovation in the private sector and for improving public services and enhancing social conditions.
>
> This raises the fundamental question of why this wealth of technology is not used and exploited to a greater extent. Many factors lead to decisions against innovative changes, including: the style of government procurement; the market structure of the public sector; the nature of the competitive process in both the public and private sectors; the lack of aggregation of both industry segments and governmental units into sizes adequate to pursue new technology; and the fact that views on technological risk taking very often lead to decisions against innovative changes.[1]

Regardless of doubts as to whether the stock of technical knowledge developed, for example, in the areas of defense, air travel, and atomic power, actually is capable of fulfilling the requirements of humanity or whether these new technologies will largely bypass the needs of the populations in the industrialized nations, the paradox of unexploited technologies remains a fact of life. If it were only a question of employing today's jobless manpower and underutilized capital in large-scale technological innovation projects, there would easily be enough practical ideas to set up extensive public works programs. But who actually finds this alternative desirable?

There is clearly an impediment to the flow of technical knowledge. Although there are many ideas for workable innovations in the theoretical world, that is, in tax-supported universities and research centers, the problem is that the practical world does not seem to be able to apply most of these new technological concepts. One of the most troublesome obstacles to innovation is, of course, that no one actually *knows* what is desirable for the future. Again, as Otto Eckstein said: "Calm times teach few lessons, but adversity is a great teacher."

In the past, the American government has made a considerable effort to spill over the technical advances made in the areas of defense, air and space flight (NASA), and atomic research (AEC) into the general economy. This forced technology transfer would certainly have benefited the Pentagon, NASA, and the AEC if it were then easier for Congress to pour billions more into defense, space, and atomic research because the consumer was also deriving some benefit from this spending. However, very few of the attempts to push these new technologies in the civilian sector have made any significant impact on the economy. The flood of knowledge from the massive government research projects did not raise the innovation rate in the private economy. Technology transfer experts have labeled this phenomenon—that is, a substantial buildup of new technical knowledge and many unsuccessful attempts to launch this particular new knowledge into the economy—the spaghetti effect.

The spaghetti effect is an inversion of the echo effect. The echo effect is an (unsuccessful) attempt to explain the lack of basic innovations because of a temporary failure of science. The spaghetti effect, on the other hand, explains the lack of innovations as the result of the inertness of the captains of industry. If you move one end of a limp piece of spaghetti, the other end will not move. This metaphor does do justice to reality. A large fund of knowledge is building up, but it is affecting actual practice at a very slow rate. It is a well-established finding of innovation research that "technology push" is an inferior way to introduce new technologies on the market; "demand pull" is a major factor for successful innovation. If this demand is lacking, the rate of innovation is low.

This paradox of unused technologies does not only characterize the past ten years' slide into the present technological stalemate, however. There are also parallels in the past; we can observe the same paradoxical situation in earlier periods of stagnation that led to technological stalemates. The same constellation of substantial technical advances in theory and very little practical application occurred in the years before the crises of 1825, 1873, and 1929.

We are making these observations using the data on basic innovations and the instigating basic inventions in Tables 4–1 to 4–4. First let us focus on the surge of innovations during the 1930s. The technological stalemate of the 1920s resulted from the lack of basic innovations before that period. However, Figure 5–3 shows that most of the essential basic inventions later to be applied in the innovative surge of the 1930s were already well known by 1925. Thus we see that the paradox of unused technologies existed even in the 1920s!

During this period, the relationship between the stockpiling of knowledge and its practical utilization was as follows:

- By the crisis year of 1929 *more than three-quarters* of the basic inventions had already occurred.
- By 1929 *less than one-quarter* of the ensuing basic innovations had been implemented.

We can see the same paradoxical relationship between a high level of accumulated knowledge and a low rate of practical application on the eve of the great crisis of 1825. The situation is shown graphically in Figure 5–2. In 1820, the fund of knowledge, as measured by the total number of unutilized basic inventions in existence at that time, comprised 80 percent of the information from which the imminent innovative surge was to draw sustenance in the years around 1825. By 1820, however, less than 20 percent of the total basic innovations that were implemented in the following years (according to Table 4–1) had in fact been put into practice. This discrepancy is astonishing given the prevailing problems with the utilization of liquid capital[2] in an economy battling a structural crisis.[3] Contemporary British economists characterized this period as a time of distress.

Actually, in the 1820-years of world economic crisis, the congestion hampering the basic innovations was not as severe as it was in the 1920s. In the last years before the collapse in 1825 certain basic innovations of great economic potential had been realized. Among these were the first production of Portland cement (1824), the first construction of a puddling furnace (1824), the first rolling of wire (1820), and the factory-processed distillation of quinine (1820). These timely technical innovations might have helped to mitigate the prevailing depressive tendency and dampen the crisis.

The world economic crisis of 1873 and the resulting depression were more intense than the crisis and depression in 1825 but less severe than the collapse in 1929. This period demonstrates the same paradoxical relationship between an initially low rate of basic innovations being implemented despite a high level of accumulated basic inventions. Figure 5–1 depicts this discrepancy. In the fields of chemistry, electrophysics, and related technologies, the swell of basic innovations occurred many years after the world economic collapse in 1873, whereas the underlying basic inventions had long before become well established in the pool of applicable knowledge.

Capital owners did not invest in basic innovation. Bismarck articulated the paradox of this situation; he is said to have jokingly suggested that in light of the difficulties that the stagnating German economy was having in productively investing the war indemnity of five million marks from the Franco-Prussian War, the French should be forced not to pay but to accept x billion marks from the Germans after the next war. Newbold, an economist writing in the depressed 1930s and trying to learn something useful from research on the depression in the 1870s, pinned responsibility for the collapse of 1873 on "the complete confusion reigning on the international gold markets."[4] This was a true but superficial diagnosis given the understanding of political economy that Bismarck had seemingly already acquired fifty years earlier.[5]

The present technological stalemate is also characterized by lagging innovative investment in the private sector. Private holders of large amounts of capital are finding no worthwhile investment opportunities. They are clearly not finding many opportunities in the multitude of innovative ideas that have been produced in large-scale government research, university institutes, and industrial research laboratories, or at least not yet.

This time, however, the awareness of the innovative possibilities lying ready or ripening in the research sector could bolster the confidence of the economic planners so that they do not panic and push us toward another world economic crisis. As they have always done in the past, the frustrated private investors may put their currently underutilized capital into these new ideas, thus compensating for the previous neglect and possibly producing a new swell of basic innovations.

The course chosen to compensate for the previous lack of basic innovations can be taken to extremes, however, as has typically been the case in the past. Hastiness may result in reckless attempts to boost the economy by introducing technologies that are of dubious social value. One word of warning concerns hastiness as such. When L. Robbins reviewed the Great Depression of the 1870s and 1880s in 1934, he stated the best way to avoid a depression is first to avoid triggering a boom.[6]

However, our first worry should be that the dammed-up force of investment energy is artificially released and directed into the nearest, quickest, and not the best innovative possibilities. In similar circumstances in the past, these were always new types of weapons and more efficient ways to produce war machinery. Not only did public investment in the war industry then crowd out private investment in innovation in peace industries, but it also induced a higher propensity of war. This is the Hitler fallacy.

THE TEMPO OF TECHNICAL PROGRESS

The thesis that technical change is being accomplished at an increasingly faster rate presently enjoys great favor. This popularity derives from the desire to believe that the innovations currently in short supply can be produced in a twinkling, or as the Keynesians believe, can be taken from the shelf at will, and thus there is no reason for serious concern. This acceleration thesis is only half-true; it is wrong in its essential point: the reliance on instant makeability of new industry.

As an example of this popular belief, read the following statement of Paul A. Samuelson, who is probably the best-known economist not only in the United States but in the whole world. In an interview with *Economic Impact* (1978/2), which is a journal produced in the neighborhood of the White House, published by the U.S. International Communications Agency, and distributed through the U.S. Diplomatic Service, Samuelson said:

> In the old days we didn't know how to make jobs. When somebody lost a job because of new technology in the buggy-producing industry, we used to pray that the automobile industry would come along and make jobs. But today we have central banks, we have monetary and fiscal policies of governments. We can, without waiting for luck, make jobs in new areas if we lose jobs because of technology in old areas.

This belief in instant makeability of new jobs, as it is in fact portraying the governing economic philosophy, clearly proves "the need for a wholesale reconstruction of the microfoundations of contemporary monetary theory" (R.W. Clower and P.W. Howitt, *Journal of Political Economy*, 1978, No. 3, p. 449);

Today one can produce minor improvement innovations with the wave of a hand in many traditional areas of technology; there are many technically realizable ideas available from these traditional areas of knowledge. However, there is not enough market demand for traditional products in an affluent society, because by definition the society has been largely satiated with them. The fact is that as the quality improvements in successive product innovations in mature markets diminish, pseudo-innovations appear in increasing numbers and in increasingly rapid succession. Indeed, there is real acceleration here.

On the other hand, what has not been grasped is that basic innovations, because they create completely new markets, today do not progress faster than in previous centuries.

Today, we have too few basic innovations in new fields of technology because the process of innovation cannot be accelerated as much as many think it can. Acceleration means rushing through research, testing, development, and experimentation of basic innovation possibilities that have been neglected during decades in the past. Only at the expense of higher risk and inferior results can this time-consuming process been shortened. Time is money, and valuable time has been wasted because the road from invention to innovation is long when it must go through uncharted industrial territory.

The tempo of technical progress in the unexplored areas of new technology is slow compared to the speed of change in more established fields of well-known technology such as aviation, synthetics, and computer technology. Today the theory of accelerating technical progress seems valid for certain types of improvement innovations, but it does not apply to basic innovations. One must examine the totality of implemented and unimplemented (or slowly implemented) innovations. There are significant differences.

The claim that there is a decreasing time span between the beginning and the execution of a change also does not gain credibility simply because it possesses impressive ancestors. When Compte formulated his first law of social dynamics in 1840, he adopted Turgot's concept of historical phases. He maintained that the theological, metaphysical, and mechanical epochs were each shorter than the previous epochs had been. The mechanical age then gave way to the electrical age, and "the stepping-up of speed from the mechanical to the instant electrical form," as Marshall McLuhan puts it, has become a daily experience for modern man.

The notion of a general acceleration in the course of history, that is, the concept of a progressively shorter period necessary to accomplish a change, was formulated as a law by Henry Adams.[7] This idea has been reflected in modern scholarly and scientific literature, particularly in the writings on the design and development of new products. Many writers claim that it has taken progressively fewer years for knowledge to travel from inventor to innovator.

Figure 6–1 provides an example of a precipitous conclusion reached after putting together a few selected observations. Figure 6–1 was taken from a work by Stamm and Wilmes,[8] which used data gathered by Brankamp. Dr. Zipse jokingly provided Henry Adams's law with a precise mathematical formula to demonstrate how easy it is to produce at least the appearance of scientific precision. He then placed the Stamm and Wilmes data along Adams's curve

$$I = A \cdot \exp(-t/T)^{-B}$$

to produce the graph presented here as in Figure 6–2. The length of innovation time calculated in years has been entered directly above the time point of discovery.[9] As a matter of fact, Dr. Zipse was not the first person to demonstrate an awareness of the nonsensical nature of the so-called Adams's law. In 1943, Winter[10] published a book pointing out how confusing Adams's thesis was. It is clear then that Adams's law is nonsense; the question now is why it makes such little sense?

The time intervals are not the problem. As Sicherl recently established,[11] the duration of a process of change can be an indicator of the distance that a backward economy must travel on the road of

Figure 6–1. Acceleration in History.

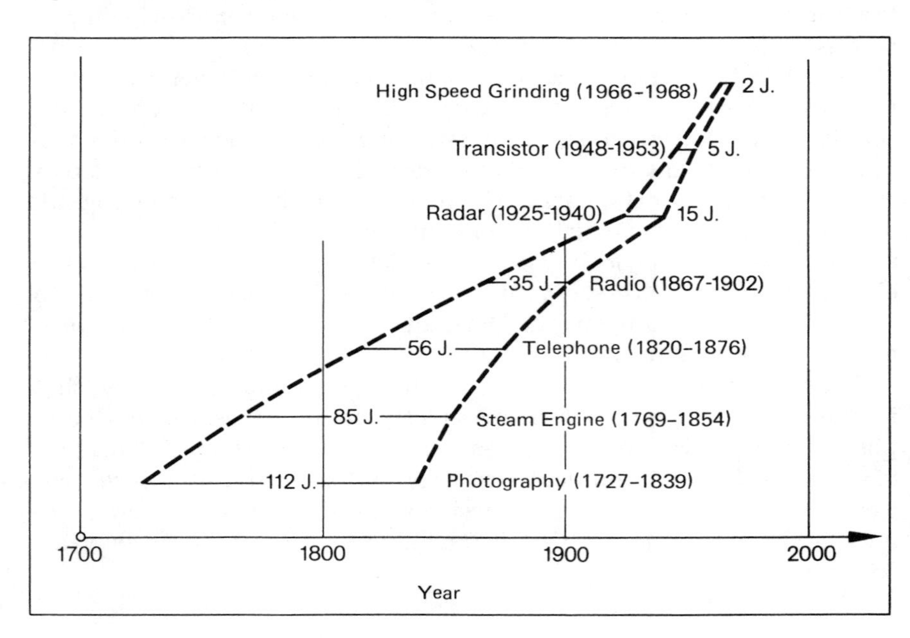

Figure 6–2. Acceleration in History.

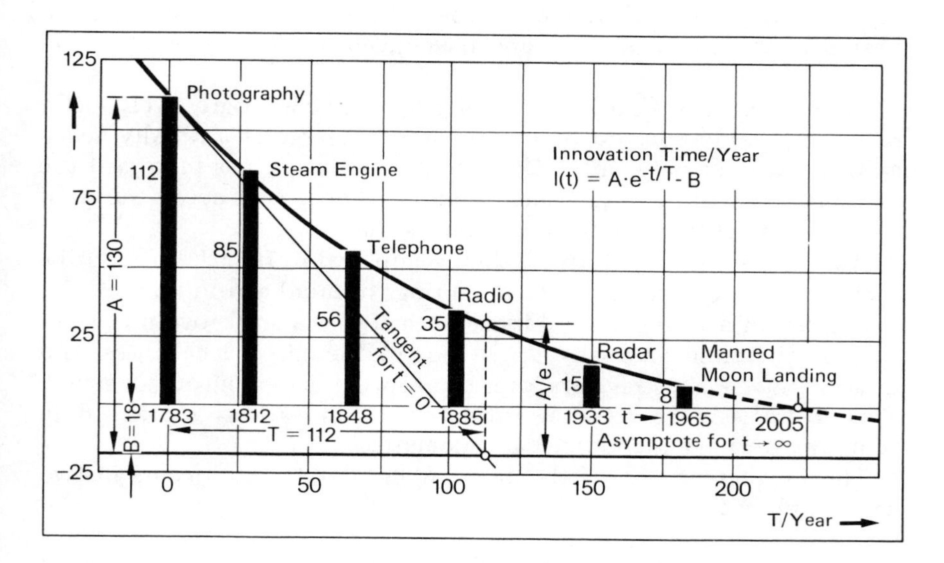

development. Similarly, the time taken by the transfer of knowledge from basic invention to basic innovation is an indicator of the differences in a changing economy at different times. The time span communicates an impression of how troublesome and tedious it must have been to change the system from its initial condition at the time of the invention to the condition it was in at the time of the innovation. *We observe long innovation periods for avant-garde innovations and shorter periods for innovations that follow familiar ground. Only pseudo-innovations occur with the wave of a hand.* This explanation clarifies the dynamics of technical progress;[12] the speed varies according to the radical quality of the innovation and the receptiveness of the economy in the various sectors.

The problem with Henry Adams's law is not in the use of time intervals, although naturally these intervals could be determined incorrectly in individual cases. The problem lies in the failure to perceive the difference in the pace of the movement of the transfer of knowledge in *different areas* and for *different types of innovations.* Certain movements demonstrate opposite tendencies from the general trend, and thus there can be no illusion of a general decrease in the time needed for structural change. The popular hope that a historical trend toward acceleration of technical progress would enable us to step in, at the right time with whatever new technology we may happen then to desire, and restrain stagnation is thus groundless. The general impression that innovation times are shortening is merely a sign that a particularly large number of pseudo-innovations has appeared in the past few years and have given us a quite distorted picture.

Basic innovations do not generally appear at a progressively faster rate. Only sometimes do basic innovation emerge at a rapidly accelerating pace, namely, after the initial retardation has increased the pressure for realization to such a degree that it produces a surge of innovations. This is but an exceptional situation.

Americans use the term, "the wagon train effect," to describe such a course of events. The name brings to mind a group of vehicles taking off in a flying start. First only a few wagons move into gear; then as the train gradually grows larger, the speed accelerates, and thus when the last stragglers join the group, it is moving at full speed. This is precisely how the swarm of basic innovations appears in an economy during a technological stalemate.

The movement out of the technological stalemate occurs in two distinct phases:

- Initially only certain individual basic innovations are realized. The mass of basic innovations is still being obstructed.
- Then, however, a swarm of basic innovations is activated *with crowding and great haste as the result.*

Thus one should be able to observe the wagon train effect in both the change in the number of basic innovations that appear and in the growing speed with which any individual basic innovation is realized. We have already found a method to observe the first aspect of the wagon train effect. Data on the changing numbers of basic innovations have been assembled in the tables in Chapter 5, and in Chapter 4 we characterized the wagon train effect as surges in the appearance of innovations.

In this chapter we want to empirically analyze the speed of the transfer of knowledge. This means we will venture into the same uncharted territory that the transfer of knowledge has led us into before. In fact, no concept has been developed for the speed of the transfer of knowledge in any area of the social sciences: "We still lack a means to measure the tempo and direction of social change."[13] We will now attempt to develop a rough measure for the speed of innovative change from the lead time covered in the transfer of knowledge from basic invention to basic innovation.

Lead Times in the Transfer of Knowledge

We would now like to describe the movement of the transfer of knowledge from basic inventions to basic innovations in specific areas, beginning with the fields of chemistry and electrophysics. With the help of Thomas Kuhn we have portrayed the scientific revolutions in these areas in Chapter 5. Lavoisier's theory of combustion was the paradigmatic breakthrough for modern chemistry, and the inventor of the Leyden jar parallels this event for electrophysics. Both areas of knowledge stem from knowledge whose origins lie in the epistemological field of the late eighteenth and early nineteenth centuries. The elan and spirit of that epoch provided a fertile base for the theoretical knowledge developing in new branches of technology, in chemistry, and electricity. Moreover, these new insights were achieved at a time when practical energies were being squandered on previously developed insights in theoretical mechanics and thermodynamics, that is, on the attempt to build a steam engine.

In Chapter 4, we outlined the course of the stream of knowledge in these areas and then recorded it as lists of basic innovations in Tables 4–2 and 4–3. These lists also include the time intervals between the basic inventions and the resulting basic innovations. The distance traveled over time is visible in Figure 5–1 from the gaps in the transfer of knowledge.

The year in which a basic invention occurred and the time elapsed before the knowledge was applied in practice are represented as points in Figures 6–3 and 6–4. One point represents both the year of the invention and the lead time necessary to achieve the innovation. The set of points for the overall transfer of knowledge shows the changes in the lead time necessary for innovations to be implemented in a particular area of knowledge. These sets of points produce a characteristic pattern somewhat like a photographic negative of the Milky Way. The points are strewn in a broad band directed downward and to the right.

The popular thesis that innovation times are continually becoming shorter is thus partly confirmed for the reduction in lead times in individual branches of knowledge. The innovation times in both chemistry and electrophysics show a clear tendency to decrease. Moreover, this reveals the leadership role of the pioneer innovations. The first basic innovations to be implemented produce an accelerating stimulus for other innovations following them.[14]

Now we will take a brief look at Figure 6–7. These clusters of points combine the data on the knowledge transfers in chemistry and electrophysics. As we have said, both areas derive from the same

Figure 6–3. Lead Times from Basic Inventions in Chemistry to Basic Innovations in the Chemical Industry in the Second Half of the Nineteenth Century.

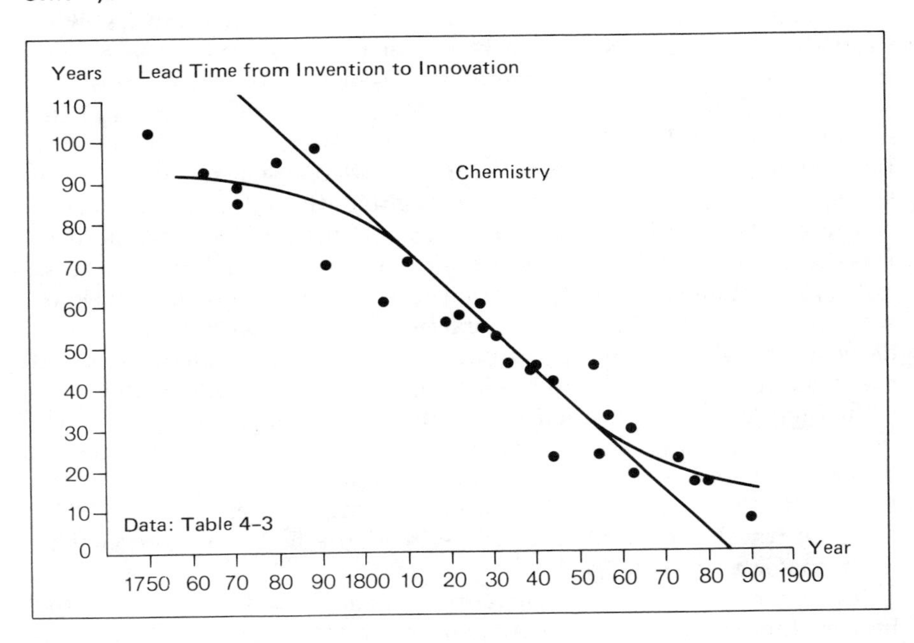

basic set of knowledge; it is also evident that the lead time patterns for both areas are the same. This means that the streams of basic inventions in both areas of science run parallel courses until they merge rapidly under the conditions of a technological stalemate (see Figure 5–1 for further illustration).

I call this double similarity *the correspondence principle in the transfer of knowledge.* It states that basic inventions in *different* areas of knowledge that, however, derive from the *same* epistemological field (different branches of general knowledge) require about the same amount of time to mature and be productive and that they become productive at nearly the same time.

The correspondence principle explains why the majority of the basic inventions conceived over a half-century or more can be realized as a surge of innovations within a few years even though their theoretical bases (inventions) are in different scientific branches. Partly this is because the scientific branches that produce practicable ideas simultaneously have generally been inspired by the same zeitgeist. Thus the individual surges of basic innovations that appeared in the years around 1825, 1880, and 1935 may be traced to differ-

Figure 6–4. Lead Times from Basic Inventions in Electrophysics to Basic Innovations in the Electrical Industry in the Second Half of the Nineteenth Century.

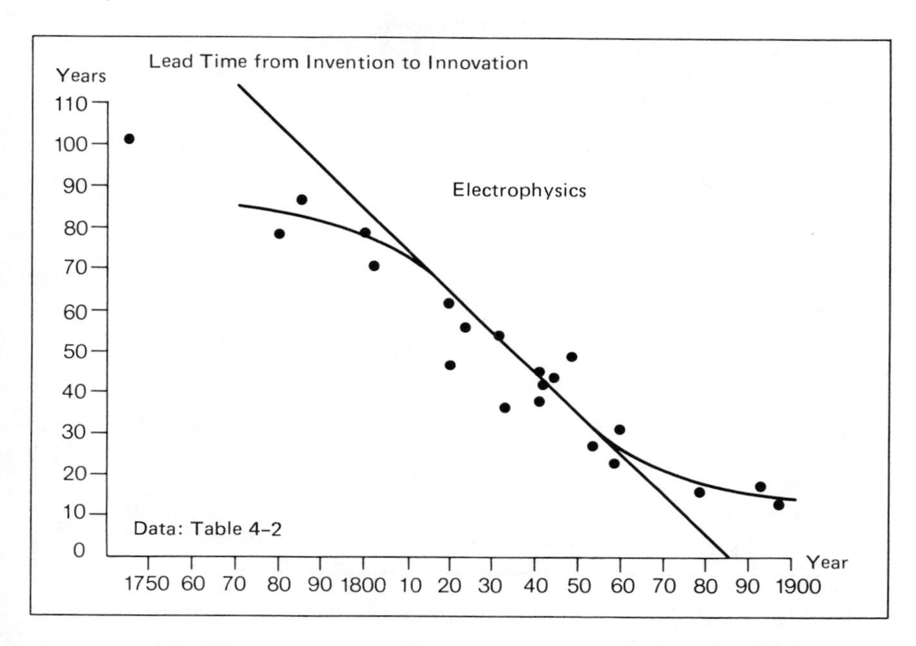

ent scholarly disciplines although they are built on the same philosophical assumptions. Only their translation into economic practice limps slowly along. The main force behind the corresponding principle, though, is the economy's initial disinterest in the scientific advances and its subsequent initial high demand for basic innovations during times of depression.

The correspondence principle also explains the striking similarities in the lead time patterns for various historical epochs given in Figures 6–3 to 6–8. Despite the fact that all of these individual basic inventions originate in different intellectual fields, the lead times are distributed in nearly identical patterns.

Figure 6–9 provides a survey of the set of lead time patterns. It shows that the tendency to move toward shorter time spans for change exists in certain areas only and only in certain periods. The duration of change in other areas may vary widely. This means to any theoretician with practical ambitions that he will have to wait an unbearably long time to see any economic results. The fate of the inventor is indeed unfortunate in the areas that lie outside of current industrial practice and concern. This is so because it could take

Figure 6–5. Lead Times from Basic Inventions to Basic Innovations in the Industrial Revolution *(around 1764)* of Brno.

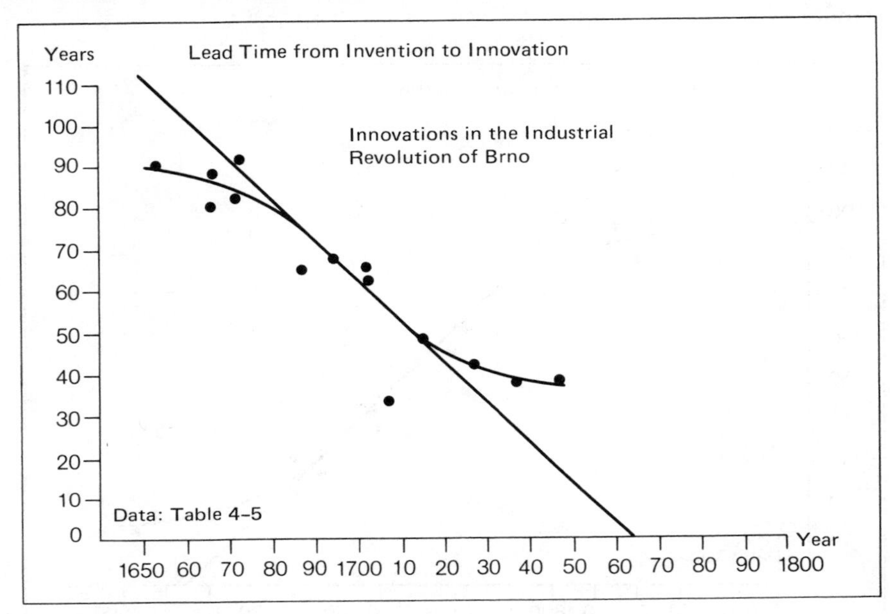

Figure 6–6. Lead Times from Basic Inventions to Basic Innovations in the First Half of the Nineteenth Century.

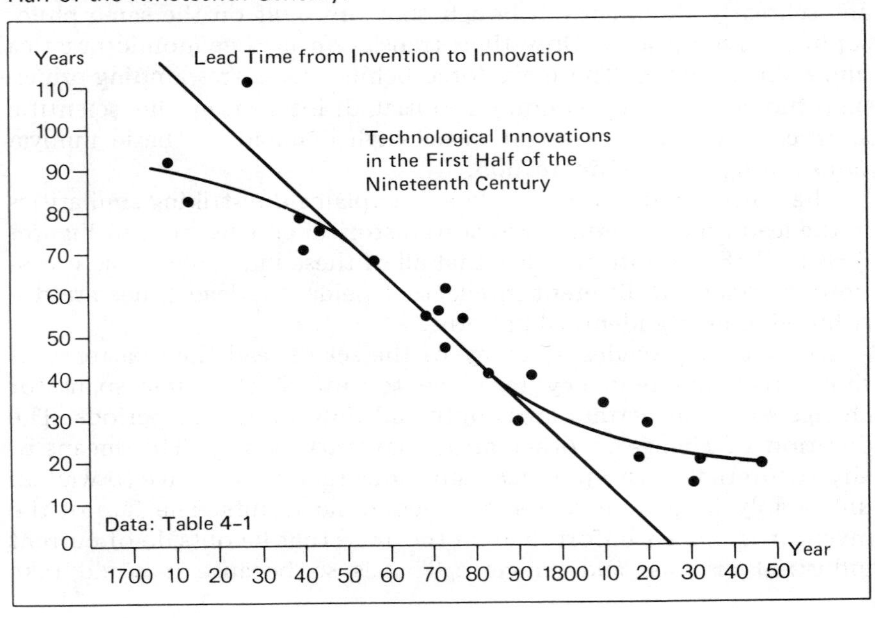

Figure 6–7. Lead Times from Basic Inventions to Basic Innovations in the Chemical and Electrotechnical Industries in the Second Half of the Nineteenth Century.

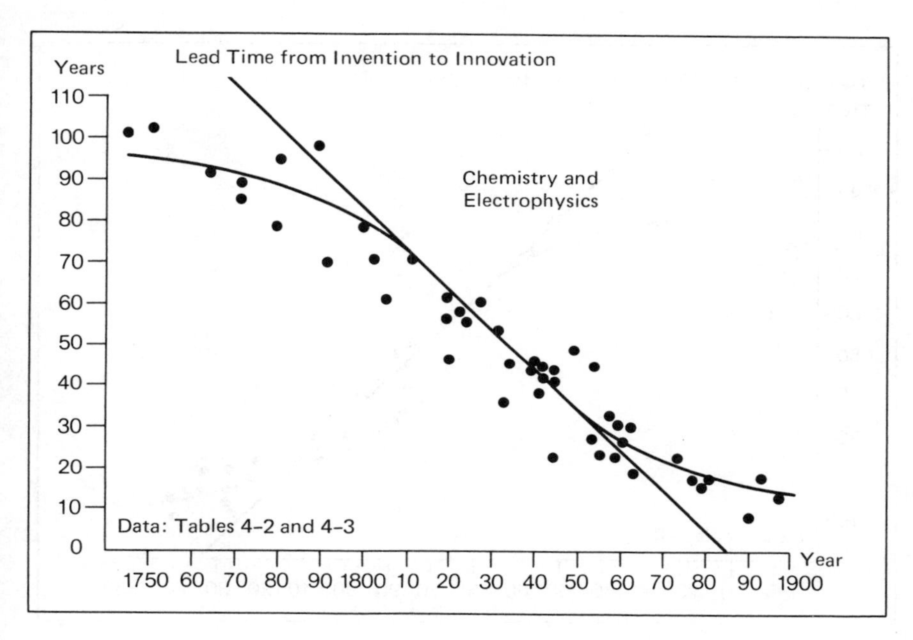

seven, eight, or nine decades from the time of the brilliant basic invention until it has made its way into economic practice. In other words, the inventor would not live to see the idea actually being applied in the real world. "Between the first and second stages there is a long road to be traveled. Sometimes the distance is so great that the road can never be covered, and the invention remains purely in the realm of ideas until the dust of time covers it, or until, under changed circumstances, it becomes favorably placed for reception."[15]

Given the lead time patterns in the knowledge transfer into basic innovations as shown in Figure 6–9, we must agree in general terms with the late Nobel Prize winner, Dennis Gabor:

> The wedding of science and technology was not achieved in one step, but only gradually, in the course of three centuries. Even in the nineteenth century it was so imperfect that, though most of the laws of electricity and magnetism were discovered and fully formulated by Faraday and Maxwell, not a single one of the electrical machines was invented in the country of the Royal Society. Nor was it perfect in other countries. Henrich Hertz produced, in 1887, the electromagnetic waves that had been implicit

Figure 6–8. Lead Times from Basic Inventions to Basic Innovations in the First Half of the Twentieth Century.

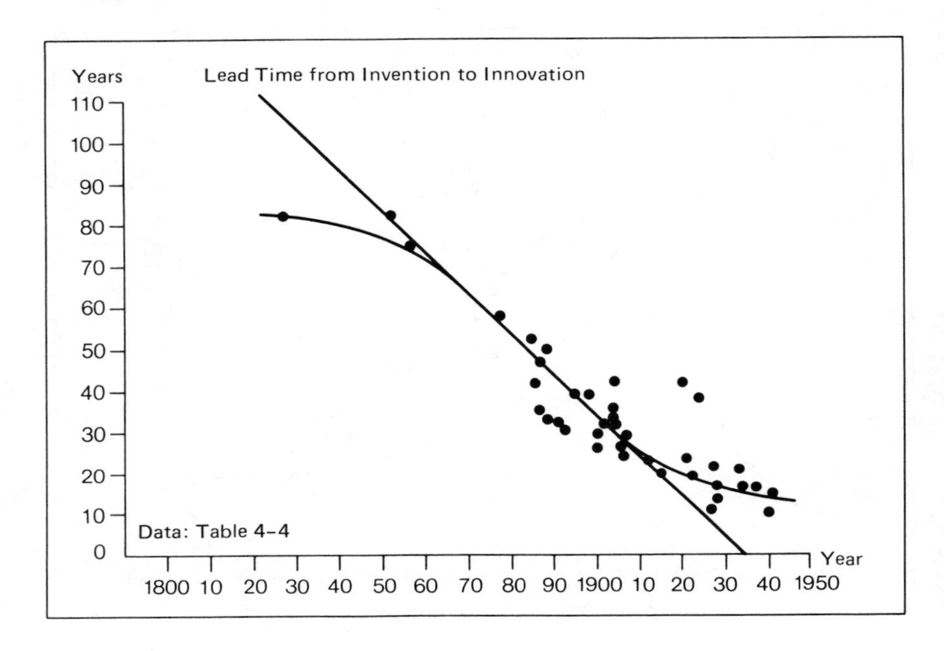

Figure 6–9. Survey of Lead Times Since the Industrial Revolution.

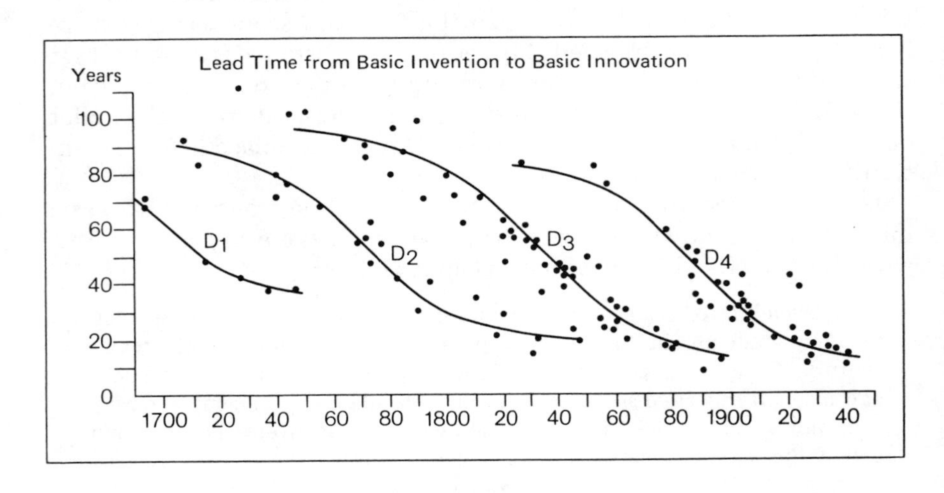

> in Maxwell's equations since 1868, but it was left to Marconi in 1896 to utilize the antenna, whose complete theory was contained in Hertz's equations. This gap of twenty to forty years between scientific discovery and technological exploitation remained typical for most of the nineteenth and early twentieth century. It was dramatically shortened only in our own times.[16]

We must disagree with the last sentence, however. First, this reduction in lead times does not, as we have seen, apply to all areas of technical scientific research; second, similar reductions in innovation time have also occurred around 1760, 1825, 1886, and 1935. One need only examine the innovative surges during those periods to establish this fact (see Figure 6–9).

Moreover, whenever the reduction in innovation time is at its height, namely, in the case of follow-up innovations in established technologies, the chances for the application of wholly new insights are radically reduced. As the inventor of holography, Dennis Gabor himself provides a perfect example of this.

In 1948, Gabor, professor at the Imperial College of Science and Technology in London, published an article in which he explained that it was possible to take photographs without using a lens and that the pictures produced were truly three-dimensional—holography. He thereupon patented the holographic process in 1948. Gabor's contribution offered a feasible solution to a problem recognized and defined as a theoretical question years before. Both the theory of radiation that contributed to the production of the first holographic pictures in 1963 and the laser beams used in the 1963 process actually derive from the equations that Einstein developed in 1917. The specific kinds of scientific knowledge that Albert Einstein initiated took a very long time from theory to practice. In 1978, after 30 years, holography, in fact, is not really on the market yet.

There can be striking differences in the innovation time required for a given technology, depending upon whether it develops within or outside of existing industries. These differences clearly refute the popular notion that the knowledge transfer proceeds more quickly today than it did in the past. For example, the basic patents for both the transistor and holography were issued in 1948. Thus, both of these technologies began simultaneously—but it is astounding how different the course of development in the individual fields has been!

Today, after 30 years, we do not have a holography industry while within ten years of the invention of the transistor, the transistor industry had developed into a multibillion dollar enterprise. The difference is not difficult to explain. From 1948 on, the transistor replaced vacuum tubes in all of their previous functions because it was more compact, much more reliable, used less energy, and was much cheaper to manufacture. It quickly established its own market, making radio tubes an absolete technology.[17] The transistor broke into an established area of technology; holography exists in no man's land. Figure 6–9 provides an overall picture of the accelerating aspects of the transfer of knowledge in different areas of science and technology that have produced structural change since the Industrial Revolution.

This survey of Figures 6–5 to 6–8 once again illustrates that innovation times have only been reduced in certain established areas of technology. The popular preconception that innovation times are shrinking does not hold true for basic inventions that break new ground and exist in an industrial no man's land. In this situation, technical progress is not a tiger (Paul Samuelson). Progress is a snail (Günter Grass).

The Speed of Transfer Processes

The time spans between basic inventions and the practical application in certain areas of technology provide the preliminary data for a rough estimate of the speed at which the surges of basic innovations occur. How can one measure the speed of innovative activity? For basic innovations, I have outlined one approach in my essay on the dynamics of technical progress mentioned earlier.[18]

The underlying concept is very simple. We have borrowed the methodology used in physics to measure the speed of movement in space and have applied it to the movement that takes place in the transfer of knowledge. This movement is a qualitative progression extending from the basic invention, i, to the corresponding basic innovation, i, an advancement that takes d_i years lead time to occur.

According to the principles of mechanics, speed is an expression that characterizes the movement of a body over spatial distance. In analogous fashion, we will express the qualitative distance from basic invention to basic innovation in the transfer of knowledge as a relative term: 100 percent. One hundred percent expresses the difference between the condition of the world before the birth of the idea (basic invention) and the world's condition after the idea has been put into practical application (basic innovation); 100 percent is a very uninformative expression. However, it is synonymous with the completion of a transfer processes, and thus it serves its purpose.

As it has no dimensions, 100 percent is a useful expression for us because we want to examine various innovative processes together in order to understand the different forms in which change occurs. Long lead times for change are expressions of the difficulties involved in introducing better alternatives to the status quo.

All that we require now are the simple rules for calculating percentages. Speed is measured by relating the distance covered to a significant chosen unit of time. I have chosen a year as the unit. The average speed of a single transfer process, i, then is equal to the reciprocal value of its duration, d_i, expressed as a percent:

$$S_i = \frac{100 \text{ percent transfer}}{d_i \text{ years lead time}}$$

This expression measures the progress in percentages, that has occurred each year in the ith transfer process between a basic invention and the corresponding basic innovation. We are assuming here that the ith transfer process advances at the same rate (s_i%) every year. This assumption is not realistic, but we will see later that it does not adversely affect our results because it is canceled out.

We now return to our preliminary problem. We want to pinpoint the trend reversal in a technological stalemate along with the sudden progress in overcoming obstacles in the transfer of knowledge. We do this by demonstrating a rapid acceleration in the speed of the transfer. For this purpose, we use the data, s_i, that is, the reciprocal value of the time span, d_i, for an individual innovation. These data are presented in Tables 4–1 to 4–5 under the column headings "Years Lead Time" and "Tempo of Change." We will compare how slowly—or quickly the earlier—and later basic innovations in each surge were implemented.

We first apply this methodology to the surge of basic innovations that the Industrial Revolution produced on the European continent.[a] The data, which derive from my Brno study (with H. Freudenberger), are listed in Table 4–5.

In Figure 6–10, the upper left corner shows the different lead times, d_i, as a series of points like the configuration in Figure 6–5.[b] These points are scattered in a lead time curve, D, whose middle section has a slope of 45 degrees. This middle section lies on a linear transformation function, which I call the transversal and which has the following characteristics. It has a −45 degree slope and intersects the time axis in that year in which all of the transfer processes whose beginning points lie on the transversal experience their breakthrough as basic innovations.

[a] I am treating the events in Brünn, the erstwhile "Manchester of Europe," as prototypical for central Europe.

[b] There is one exception; the point (1704, 36) was left out because it lies far outside the overall point distribution. In terms of the dynamic revealed in Figure 6–9, one can easily identify this point as a straggler from an earlier swell of basic innovations (see Figure 6–9 for further clarification).

Figure 6–10. Lead Time and Speed of Innovative Change.

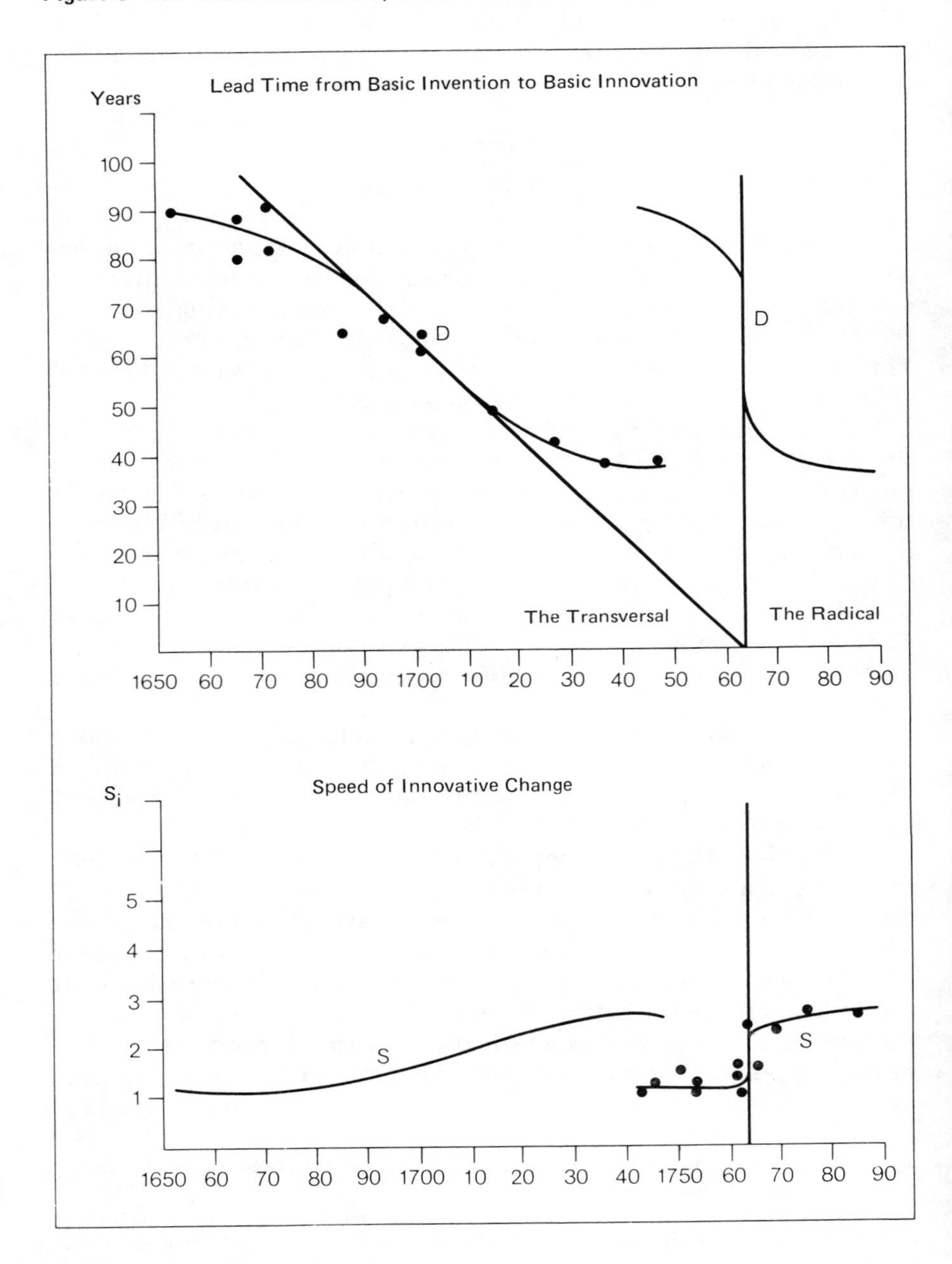

The perpendicular line rising from this point in the lead time is called the radical, and the year itself is called the radical year, that is, the year in which the surge of innovations produced the most basic innovations. 1825, 1886, and 1935 were radical years.

Let us now examine the other distribution of points in the lower right portion of Figure 6–10. These points represent the average speeds of the individual transfer processes during the years in which the related basic innovations occurred. These points cluster along a tempo curve, S, whose middle section runs perpendicularly along the radical.

One can derive the tempo curve S from the lead time curve D in two ways. One method involves transposing D across the radical, producing D', and then flipping the curve D' over (division) so that $S = 100\%/D'$. The other method involves division first; $S' = 100\%/D$ and then the transposition of S', which produces S.

How is the tempo curve S useful? In relative terms, it depicts the speed at which the gap between theory and practice is being bridged. Up to the radical year, the point where the radical stands on the time axis, the average speed of the transfer of knowledge is slow. The S-curve is flat. At the radical itself, the tempo curve swings into regions where progress occurs at higher speeds. This means that innovative activity is receiving a strong accelerative impetus, which in the case of the Industrial Revolution brought a sudden increase in the speed of change with it so that in 1764 the rate of progress had more than doubled.

The S-curve in Figure 6–10 depicts this sudden doubling of the average speed of change in radical year 1764. The acceleration by more than a factor of two that is expressed there should be viewed as the most conservative estimate of the actual acceleration in the final spurt of the transfer of knowledge. This is because the data for S_i are average values and therefore include all of the slowdowns that occurred in the early phase of the transfer. This means that the assumption that the progress from basic invention to basic innovation occurs at the same rate every year causes an underestimation of actual values. Since we are only trying to establish that a tempo upswing actually occurs (ordinal comparison in cardinal numbers), of the degree of its acceleration does not pose a problem.

As we are interested in ordinal comparison of indicators for the speed of change at points in time, we are actually free to systematically increase or decrease the cardinal values of the tempo curves' ordinates (stretching vertically). This does not change the position of the curves on the time axis; it only changes the visual impression of the radical quality of the acceleration in innovative activity. We have taken advantage of this freedom in Figure 6–12 in order to decrease rather than intensify the impression of a radical adjustment in activity level.

To expose further the popular misconception of a general acceleration of technical progress, we now determine the average speed of basic innovation in the three substantial surges of basic innovations since 1800. Thus, with the proviso that the lead time curve D lies on the transversal, we have analyzed the series of points in Figures 6–5 to 6–8 using a graphic regression analysis. We end up with the consecutive lead time curves D_1, D_2, D_3, and D_4 in Figure 6–9. These duration curves have been transferred to the upper graph in Figure 6–11, whereby the four transversals that appear in the radical years 1764, 1825, 1886, and 1935, respectively, have been superimposed on one another to produce a single trend line.

The lead time curves on this trend line shift farther toward the lower right corner the more recent the data is on the particular transfer of knowledge. Thus, there is some grain of salt in the notion of a general acceleration in the rate of change.

Figure 6–11 shows the derivation of tempo curves, S_1, S_2, S_3, and S_4, from the lead time curves D_1, D_2, D_3, and D_4. One can calculate one's rate of speed given the distance between two milestones and the amount of time it takes to travel between them by using reciprocal values; one can also derive tempo curves from duration curves using the same logic. The process involves shifting the angled transversal in the top portion of Figure 6–11 so that it is superimposed upon the perpendicular radical in the lower graph.

The consecutive innovative surges that we have observed clearly exhibit a very unique and repetitive change in tempo. The starting points for each tempo curve were at the same low levels of innovative activity. The initial phase lasted a long time; this was true in every era.

Looking at the shape of the individual S-curves, it is clear that once the surge of basic innovations was fully functional, follow-up innovations appeared in the new areas at an accelerating rate of speed. Now, if we compare the height of the four S-curves, we see these curves indicating acceleration of advance of follow-up innovations in the surges of basic innovations. From this comparison it also follows that the theory of acceleration of technical progress that we have seen is applicable only to individual sectors is also only applicable to some periods in the flow of time. In the last two decades, for example, there certainly was a speedup in the emergence of improvement innovations, but the tempo of basic innovations was probably as low as around 1750.

Figure 6–11. Lead Time Decrease and Speed Increase at Various Stages in Economic History.

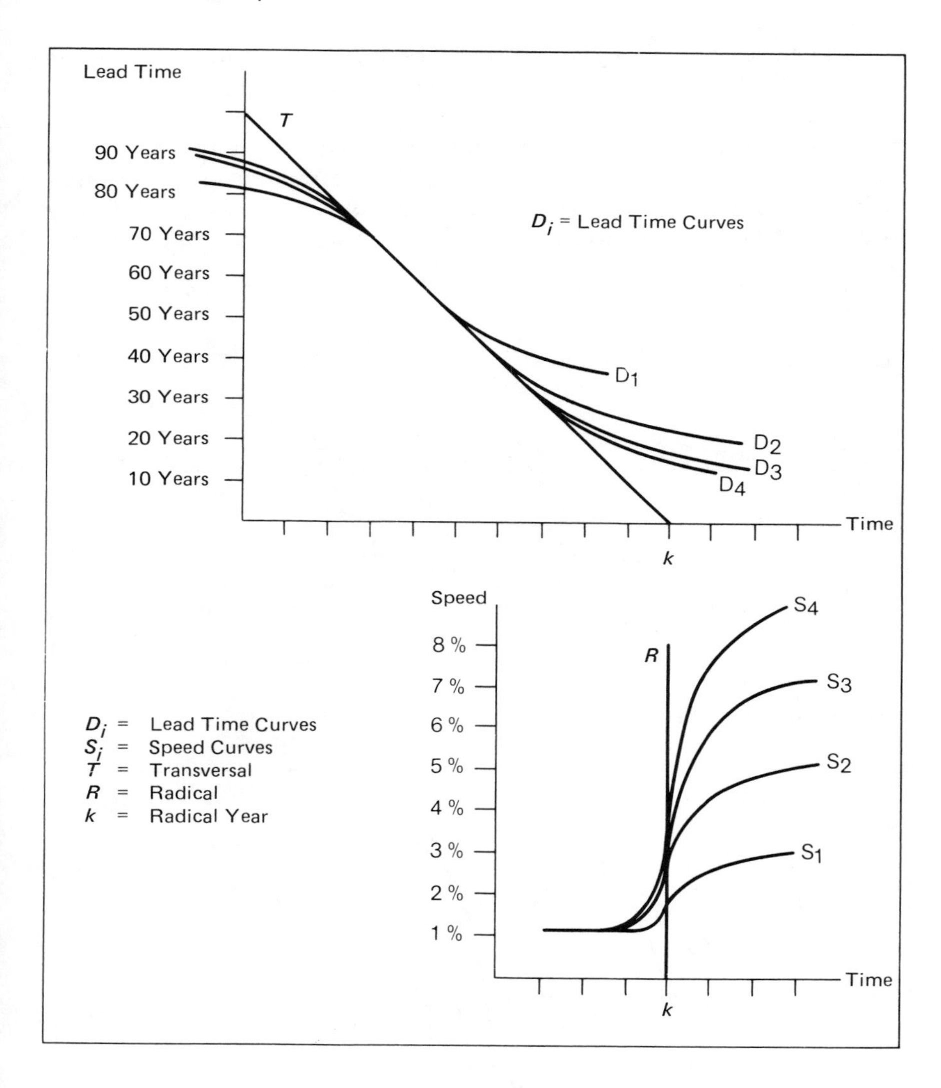

During a technological stalemate eventually the tempo of basic innovations shifts into higher gear as a new thrust of basic innovations occurs and new ideas are accommodated for the first time in the creation of novel industrial branches. A tempo curve demonstrates both the faltering speed at the beginning of an innovative surge and the increasing acceleration of change as time passes. Figure 6–12 shows the tempo of change in the successive surges of basic innovations.

Earlier we characterized the herd behavior of innovators as the wagon train effect because an innovative surge is marked by both clustering and acceleration. How do these two aspects relate? We can show their interaction by combining our figures on the tempo of innovative change with the figures on the frequency of basic innovations. These two phenomena must occur simultaneously in any surge of basic innovations. Figure 6–13 graphically confirms this

Figure 6–12. The Pulsating Acceleration of Innovative Change in Economic History.

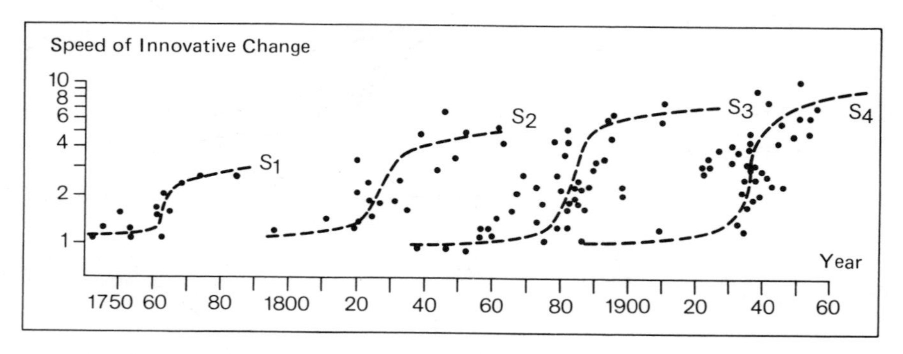

Figure 6–13. Correlation of Frequency and Speed of Change.

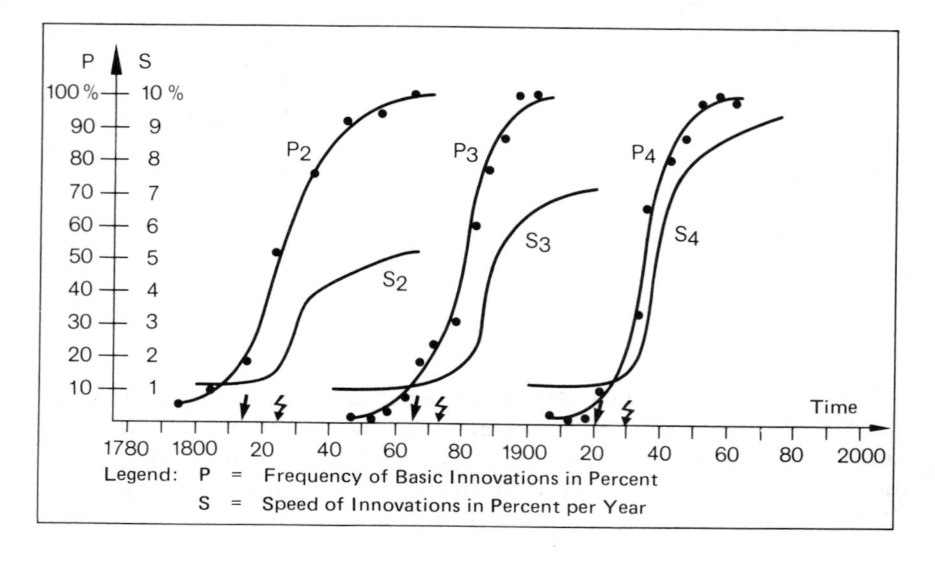

interdependence by means of a high correlation. The curves P_2, P_3, and P_4 represent the accumulated frequency with which basic innovations have occurred, according to Tables 4–1 to 4–4. A comparison of these curves with the tempo curves S_2, S_3, and S_4 reveals that both sets of curves indicate strong upswings in the same periods.

This correlation between the relative frequency and average speed of basic innovations comes into even clearer focus when we normalize the frequency figures (producing the first derivative of the curves P_2, P_3, and P_4.) These bell-shaped curves f_2, f_3, and f_4 are depicted in the lower portion of Figure 6–14. A comparison of these relative frequencies with the absolute frequencies in Figure 4–1 shows the effect of this normalization.

The characteristics of innovative surges emerge most clearly in Figure 6–14. *In the radical years 1825, 1886, and 1935 the relative frequency of innovation and the acceleration of the innovative tempo are both at their peaks.*

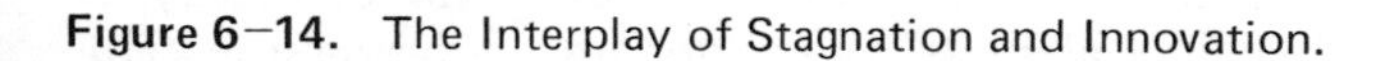

Figure 6–14. The Interplay of Stagnation and Innovation.

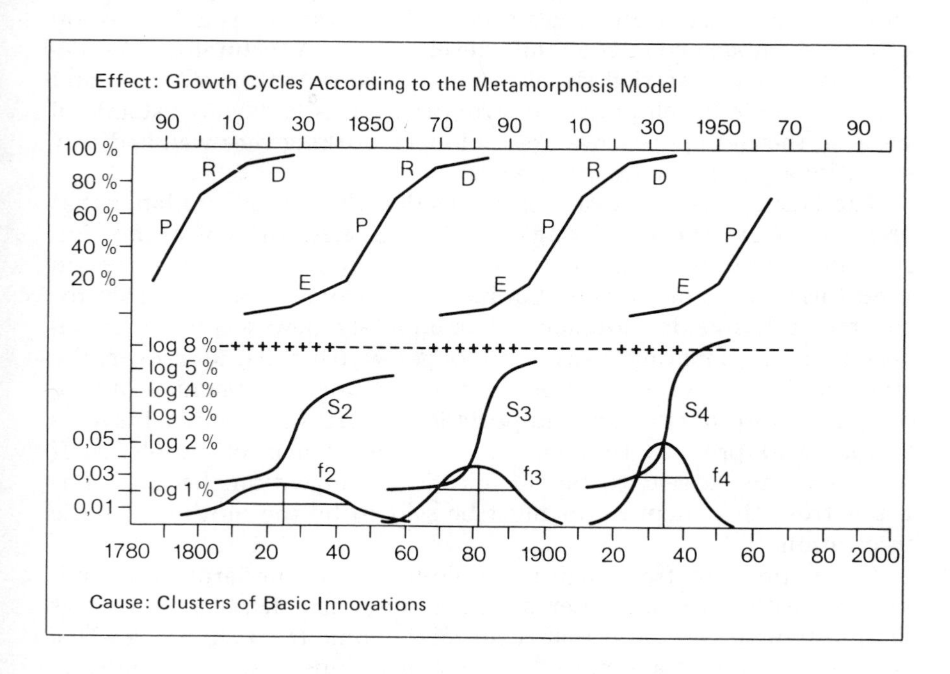

THE INTERPLAY OF STAGNATION AND INNOVATION

Figure 6–14 provides a 200-year overview of the periodic return of the wagon train effect. The succession of upheavals, innovative surges, or industrial revolutions, as some call these phenomena, can only be understood when viewed in the context of the economic conditions prevailing at the time.

To depict the socioeconomic climate of any given period during the past 200 years, we fall back on the metamorphosis model that we developed near the end of Chapter 2. It presents the long-term economic process as a succession of phases—prosperity (P), recession (R), depression (D), and recovery (E). These trends are reproduced schematically in Figure 6–14.

From these trends we can establish that phases of prosperity in the long-term economic process give way to recession when the rate at which basic innovations appear has remained at a low level for a long enough time for stagnation to gradually gain the upper hand over the invigorating impulse in economic life. The economy is caught in a technological stalemate and slides into a depression. The formula that stagnation equals a lack of innovations is a description of a gross imbalance in economic development. A technological stalemate is a very unsettled situation; industrial growth is sluggish, and socioeconomic development in structural terms comes to a standstill because the old approaches are no longer working very well and new evolutionary approaches must be developed.

The lack of basic innovations means that the capital and labor that are freed from the stagnating branches of the economy cannot immediately find new opportunities to be put to use. The recession then turns into a depression. Capital now becomes available again for the research and development of completely new technologies; investment in stagnating areas is no longer worthwhile. Moreover, the state, under heavy pressure because of the unemployment problem, develops job programs and supports the efforts of inventive business people who promise to help with the labor market problems. If necessity can be said to be the mother of invention, because it is often true, then depression must be said to be the mother of basic innovation.

Under pressure from high unemployment and underutilized capital, opposition to and reservations about untried, risky new ideas disappear with the sense that relief might come from anywhere. This situation produces a surge of basic innovations that ends the stagnation gripping the economy in the technological stalemate and ushers in the recovery phase.

Figure 6–14 shows when and at what tempo innovative surges have occurred—around 1764, around 1825, around 1886, and around 1935. Each time the momentum has increased as the process moved forward. One can easily see how this sudden demand for innovative opportunities leads to hastiness. The wagon train effect has an electrifying appeal that historically has been oriented primarily toward new technology that can be rapidly implemented rather than toward social innovations. As a general rule, the more recent the turnaround, the more pronounced the acceleration in the emergence of new technologies.

In this haste, slowly developing innovation possibilities are at a disadvantage; beneficial inventions require time to be developed. Haste intensifies the competition between alternative innovation possibilities within a given area of technology and it intensifies the rivalry between technical and nontechnical innovations. Thus, hastiness results in a biased selection of innovations: Better but slower innovations might lose out against quicker innovations.

Today, the forces for a new swell of technological innovations are quietly gathering strength under the conditions of the technological stalemate. There is some danger that the business climate, despite expansionary policies—or perhaps because of errors in the expansionary policies—will worsen to the point where serious reservations about the negative side effects of new technologies are shunted aside. Concern about unsafe types of atomic reactors, excessively expensive public transportation systems, or environmental standards might then be disregarded.

This priority shifting is already in evidence today in several countries. For example, the desire of the majority of the German population for social innovations, which caused them to vote for the social-liberal coalition, has already been given a low priority officially in the German government's list of priorities. In the past, refusal to consider social innovations and emphasis on new technological developments has always prevailed during restorative periods, which normally come only shortly after the collapse of liberal movements in times of stagnation. These circumstances undoubtedly had a decisive influence on the form of the innovations that were selected from the alternatives then available.

The market mechanism has always suffered from a fundamental weakness; it tends to neglect necessary long-term precautions such as investments in basic innovations because it puts short-range profits ahead of long-range welfare goals. The market process tends first to neglect basic innovations and then to produce clusters of them with strong support from the government. In its haste, the market process may select new technological developments that point in suboptimal directions. Given today's technological forces, these developments may turn out to be a critical problem for humankind. This is the fundamental issue involved in the notion of a stalemate in technology. We may succeed in stabilizing the economy and suffer from technological suffocation.

SUMMARY

This is the drama of the technological stalemate played out repeatedly every two generations. At first the economy does not produce enough basic innovations outside of the satiated markets and crisis-ridden industries; it therefore falls into such a depressed state that large amounts of capital are underutilized and many people can no longer find employment in industry. The lack of basic innovations is clearly visible on the eve of every world economic crisis; it was apparent in the years before 1825, before 1873, and before 1929. Then, under the pressure of those crises with the help of the government, the economy implemented a massive swell of basic innovations (new consumer goods, means of transportation, materials, and production processes) from which a number of new industries were created.

These substantial technical and organizational achievements, however, also share the stage with too expensive and too risky innovations and many varieties of "major contributions to minor needs," as the Supreme Court once put it. There are reasons to believe that in the haste of events more important innovations are being pushed aside by projects that offer little more than instant marketability. If there happened to be too many of these, the economy has no more resources available for the better innovations that come later, or the population has no more tolerance for any type of innovation, because people can only support and cope with a certain number of changes at a given time. Hence, speed is the enemy of quality.

Many economists and economic policymakers today believe in a general acceleration of innovative change throughout history—and particularly at present. This belief implies that there exists today something like a plentiful treasure cabinet which has miraculous properties: it provides the investor with easy access to precious innovation possibilities whenever he needs them, and it is always rapidly replenished through the "Process of Progress." This belief in the instant makeability of new types of investment and new jobs is ill-founded; it is falsified by our empirical analysis of the process of basic innovation.

Our conclusion, therefor, is this: The speed of basic innovative change is not accelerated in history, and it is generally lower than is often assumed, and it does not offer easy escape from a low rate of innovative investment in times of a technological stalemate. Rather, when the speed of basic innovative change indeed is exceptionally accelerated by market forces or government intervention, then the unintended side-effects of hastiness may ultimately outweigh the short-term gains.

We have thus touched upon two problems that are closely intertwined with innovative activity in the Western economy. They are the stability problem in economic growth and the selection problem in economic change. We cannot remain indifferent to the possibility that the market mechanism will again fail to counterbalance the stagnation in traditional industries and thus allow many branches of the economy to slide into crisis. Moreover, we have reason to be concerned about the chance that when the economy does recover, it may acquire so much ballast and waste material along with the technological innovations that new difficulties may soon begin to make inroads on its health.

We can offer no pat solution for the problem of selection in economic change; we can only point out its existence, its current relevance, and the danger that governments under the pressure from unemployment and inefficient use of capital during a depression will set aside the public's reservations about unsafe technologies. The creation of new jobs and new uses for idle capital was always the overriding priority during the periods of economic crisis. Given the destructive forces that our currently new technologies can unleash, this selection problem is more serious today than ever before.

To solve the selection problem, the appraisal of technological basic innovations must occur in a more thoughtful manner. This takes time. Time is precisely what was lacking in the great depressions after 1820, 1873, and 1929, and this lack of time for reflection today is what poses a real danger for the future of our children and grandchildren.

REFERENCES

1. M.D. Robbins, *Federal Incentives for Innovation* (Denver: Denver Research Institute, Report C790 to the National Science Foundation, November 1973), p. 4.

2. Knut Borchardt, "Zur Frage des Kapitlmangels in der Ersten Hälfe des 19 Jahrhunderts," *Jahrbücher für Nationalökonomie und Statistik* 173 (1961), pp. 401–421.

3. G. Mensch (with H. Freudenberger), *Von der Provinzstadt zur Industrieregion (Brünnstudie)* (Göttingen: 1975), pp. 45–46.

4. J.T.W. Newbold, "The Beginnings of the World Crisis, 1873–1896," *Economic History*, 2 (1932), p. 437.

5. H. Rosenberg, *Große Depression und Bismarckzeit* (Berlin: 1967).

6. L. Robbins, *The Great Depression* (London: 1934), p. 171.

7. H. Adams, "The Rule of Phase Applied to History," in: E. Stevenson, *A. Henry Adams Reader* (Garden City, N.Y.: Doubleday, 1958).

8. K. Stamm and P.G. Willmes, "Produktplannung und Produktentwicklung als zentrale Stabsaufgabe," *Hoesch-Estel* 2 (1973), pp. 41–50.

9. H.W. Zipse, "Beherrschung der Dynamik mehrstufiger Innovationsprozesse," *Hoesch-Estel*, 2 (1973), pp. 51–58.

10. Y. Winters, "Henry Adams, or the Creation of Confusion: The Anatomy of Nonsense," Norfolk 1943.

11. P. Sicherl, "Time-Distance as a Dynamic Measure of Disparities in Social and Economic Development," *Kyklos*, XXVI (1973), pp. 559–578.

12. G. Mensch, "Zur Dynamik des technischen Fortschritts," *Zeitschrift für Betriebswirtschaft*, 41 (1971), pp. 295–314.

13. R. Dahrendorf, *Pfade aus Utopia* (Munich: 1967), p. 323.

14. G. Mensch, "Zur Dynamik des technischen Fortschritts," op. cit.

15. M.T. Hodgen, *Change and History* (New York: 1952), p. 45.

16. D. Gabor, *Innovations; Scientific, Technological and Social* (London: 1970), p. 5.

17. N. Lindgren, "A New Process Gropes for a Market," in: D. Allison (ed.), *Dealing with Technological Change* (Princeton: Princeton University Press, 1971), p. 133.

18. G. Mensch, "Zur Dynamik des technischen Fortschritts," op. cit.

 *Part III*

Challenges and Chances

"New times demand new measures and new men" (James Russell Lowell, A glance behind the curtain)

". . . the sort of man we need in public life. One who is able to limit theory by practice yet enlighten practice by theory" (Gore Vidal, 1876)

During growth phases, the industrial economy tends to concentrate, and it concentrates in areas of activity with the highest potential for *commercial* success. From a profit perspective this mode of behavior is completely rational; it has its less attractive aspects: Concentration breeds inflexibility. The established industries tend to neglect or to start too late the exploration of new industrial areas that, in the longer run, are most promising for the capital and labor that will eventually be freed from overgrown, widely exploited areas of activity. Investment in invention and in search for basic innovations usually does not promise such a swift or certain return as investment in the expansion of mass production of commonly used goods. This is true at least in situations where the limits to market growth are not evident. Generally, "the incentive to invent is less under monopolistic than under competitive conditions, but even in the latter case it will be less than is socially desirable," wrote Kenneth Arrow, the leading welfare economist. Specifically, he suggests that "a bias against major inventions" does exist. Inertia is the intimidator of invention and innovation.

Major American corporations such as Dupont and General Electric spend approximately 90 percent of their research and development budget "in defense of existing lines of business," said their research and development managers in an informal meeting. Thus, despite the fact that research and development expenditures comprise 2.3 percent of the American GNP, there is relatively little money available in industry to investigate completely new avenues of industry. Basic innovations, the evolutionary chances of transition, receive little support in a milieu of concern for tradition. If the forces of tradition stifle basic innovation, the process of creative destruction will sooner or later have its way. In such periods of stalemate in technology in the history of modern civilization the industrial evolution always eroded or circumvented adverse government power and market power by means of a rapid succession of major innovations.

Industry always has to set priorities in its production program. It must always concentrate on high yields; that is, it will always prefer the group of products that corresponds to the pattern of high consumer spending whether this heavy spending is due to popular taste, material constraints, or government regulation. The degree of concentration on the suppliers' side increases as consumer demand tends to concentrate on the markets for the most favored goods and services. An obvious cause and effect of concentration in industry is the acceptance of more and more product standards and voluntary rules of market conduct, making real variations in the goods and services offered on the concentrated markets increasingly less significant. This tendency is self-perpetuating. It sacrifices a high rate of product quality improvement in order to gain economics to scale in mass production. As a result, quantity wins over quality as the market structure becomes increasingly concentrated both on the supply and the demand sides.

The expanding industries then neglect investment in basic invention, and they delay or even avoid implementing basic innovations. They would rather stick to the market segments that they dominate. Market power is the prerequisite of inertia. In the long run, however, readjustment in supply become unavoidable because of shifts in purchasing behavior caused by the satiation of demand in many traditional areas and the evolution of unfulfilled needs in other areas. In the 1960s the dominant industries tended to substitute pseudo-innovations for feasible improvement innovations that could have revitalized the sluggish demand. Moreover, the incontrovertible market leaders were able to cut off quality competition of smaller or of foreign firms, thus gaining a free hand in raising prices within their market segments despite noticeable stagnation of total demand growth. This paradox explains stagflation and can best be explained by increasing concentration in the late 1960s in many branches of industry. Many studies on concentration and market power that have been conducted in recent years discovered an alarming increase in the degree of industrial concentration in the economies of nearly all Western countries, and this concentration was by no means confined to only those sectors where monopoly privileges created prohibitive barriers to entry of competition. Nor was this increasing concentration limited to sectors where market power is based upon natural monopolies that owners of water supply systems, oil pipe lines, telephone networks, rails, or power stations do in fact enjoy. Nor was concentration restricted to lines of commerce where government regulation helped preferred suppliers to put competitors at a severe disadvantage. A wide range of branches of industry suffered from concentration as a result of proprietary rights (patents, etc.) on advanced technologies, rights that establish discretion over competitors who depend upon these rights being shared. To the extent to which such rights are guaranteed by government, as is the case with patents, market power is simply a derivative of government power.

In any type of human activity, it is rational to concentrate on what is essential, and this applies to both individuals and communities. Self-restriction increases one's real powers and performance. "In der Beschränkung zeigt sich erst der Meister," said Goethe. However, people can easily appear limited if self-induced restrictions become exaggerated. The same problem occurs when capital and labor are increasingly concentrated within an artificially limited spectrum of quality and specialization, thus unduly narrowing the spectrum and constricting the flexibility of the economy. The limits to growth in the fully expanded branches of the economy become chains whose stranglehold only tightens if employees and employers fight against stagnation by raising prices and wages. In this way the problem of adaptation is shifted onto the consumer. But many consumers will simply leave the market. And if many consumers leave many markets, the total volume of demand in the entire economy decreases.

In an affluent society, additional effective demand can be mobilized only by overcoming the qualitative discrepancies between industrial supply and people's real needs. Macroeconomically, where aggregate demand is the problem, it does not matter whether some firms put more capital into more efficient production technology. The effect of such investment usually is only to reduce further the level of capacity utilization in all firms and to bind more capital into a larger number of capital-intensive enterprises. It also does not matter if branches of the economy take refuge in higher prices. Raising prices only frightens off the potential purchaser even more. As the need structure of populations shifts increasingly with the degree of affluence, affluence can be maintained only if the production structure adjusts accordingly.

In Chapters 2 and 3, we showed that overconcentration in the leading industries puts a price spiral into gear. However, as the consumers' reaction demonstrates, concentration also reinforces the stagnation trend. In short, overconcentration causes markets to shrink and at the same time increases the price of less valued services. Once stagflation moves into gear, it becomes self-instigatory; stagnation feeds inflation and inflation breeds stagnation. The period around the end of 1974 and the beginning of 1975 provides two particularly appropriate examples. First, despite the reduced demand for Volkswagens in 1974 (the Volkswagen Company suffered combined losses of 100 million marks in its six domestic plants and 86,000 of the 110,000 workers had to work shorter hours) the company raised its prices three times within a year. On March 11, 1974, there was a 6.35 percent increase, on May 13, 1974, a 6 percent increase, and at the end of 1974 another 3.5 percent increase. Second, while the number of passengers flying in and out of Berlin on Pan American, Air France, and British Airways continued to drop (down 12 percent in 1972, 15.4 percent in 1973, and 12.4 percent in 1974), the airlines continued to demand higher and higher fares. They also managed to raise their prices three times within a year. There was a 9.5 percent increase on February 15, 1974, a 10.4 percent increase on June 4, 1974, and an 8 percent increase in early 1975. Since that time, price increases driving consumers out of the market have continued, and in most countries government-regulated prices increased even more than pure market prices did. One consequence of higher than average price increases in the regulated areas, where consumer demand tends to be inelastic, has been an alarming growth of consumer indebtedness.

Price manipulation despite shrinking demand clearly demonstrates the alliance of labor market power and capital market power in many branches of the economy. This is especially true in the government-regulated industries, which became accustomed to considering the wishes of the consumer only incidentally or to ignore them completely when formulating their economic plans. This habit of ignoring the consumer developed during the boom period when most markets were sellers' markets because demand was generally unsatiated. At that time, the desires of the producers (employees and employers) and of the private, industrial, and government consumers were still in harmony in most markets. As the powerful and privileged suppliers succeeded in the late 1960s in shaping the market structure mainly to their wishes, stagflation resulted. Now, during the technological stalemate, the desires of the producers and the consumers diverge in many markets, and consumers no longer behave according to the expectations of the suppliers. These discrepancies destabilize the economy worldwide, and this destabilization creates a new type of macroeconomic problem: Restructuring.

The problem is how to cope with the dilemma of tradition and transition. We depend on both to a certain extent. This problem has become critical again in the present phase of technological stalemate as power and inertia in stagnant areas have immobilized the process of structural adjustment and innovation.

In the evolution of plant and animal life, in the ecology as in the economy, crisis results from a disequilibrium of the relevant forces and interests. Unstable passages in the industrial evolution are no excepion from this principle. Vested interests produce critical situations. A conflict of interest causes economic policy to exaggerate rather than to calm the trouble since it is almost always geared to serve the strongest interest. In modern society, economic roles become confused. The majority of voters used to be either producers or consumers. Today an increasing segment of the population (young voters, retired people, and government workers) identifies more strongly with consumer interests than with employer or employee concerns. This is where the shoe begins to pinch in politics. People react to government partiality and to market power as it is only a derivative of governmental power. Abuses in the government-regulated public health system; exorbitant rents in state-subsidized housing; unfair subsidies to farmers caused by politically manipulated product prices; value-added taxes on inflated industrial prices; devaluation of savings through state-regulated monetary policy; and high direct taxes that are being put into dubious expenditures are all factors that hurt the majority of consumers of goods and services. And they tend to blame the government for the market failures. This has been demonstrated by the voting behavior in almost all of the Western countries. Apart from splinter groups, the typical voting choice in Western industrial societies is between so-called conservatives and so-called progressives. In the last five to ten years in most countries, the electorate has tended to vote against the ruling party, whether it is conservative or labor, which has resulted in a pattern similar to betting black or red on the roulette table. In the final counting of votes, the tally was fifty-fifty. This has paralyzed effective democracy in many Western nations, parliamentary stalemate and ungovernability being the obvious effects.

Dissatisfaction with price and quality of the available supply translates into discontent with power. From the consumer's standpoint, many useful innovations are needed in the commodity-producing sector of the economy; this need is even more pressing in the service industries. Moreover, in the government-regulated industries, many of the currently sanctioned market practices that exist to the producer's benefit and the consumer's detriment ask for a change. Assuming that changes conformed to a new quality consciousness as well as to shifted consumer needs, the implementation of these basic service and social innovations would create so many jobs and useful occupations that party policies would not have to be concerned with producers' interests to the same extent that they currently are. Structural instability in the economy has at all times been the major prerequisite of war, revolution, and civil disorder as well as for authoritarian regimes. But there is still plenty of room for improving the market economy and our form of government administration, provided that the necessary adjustments are not delayed so much that the evolutionary vacuum obtains revolutionary proportions.

"Since the future is hidden from us till it arrives," wrote Arnold Toynbee in *Change and Habit*, "we have to look to the past for light on it." It is my reading of the situation that the industrial world has again become structurally ready for a breakthrough of major innovations. When in similar situations in the past a swarm of the basic innovations had restructured the economy so that it better served the shifted needs of the population, in retrospect it invited many to perceive of such transitions as industrial revolutions. A number of books have been written on *the* Industrial Revolution, and on the first, second, and third industrial revolution, and a considerable disagreement exists among the authors as to which one ought to carry which number.

The question is therefore—are we again heading for an industrial revolution? Given the degree of life's penetration by technology, this would mean a "hard" transition into a state of hyperindustrialization. Alternatively, what are the chances of a "soft" transition into some form of postindustrial society?

To quote Robert Merton;[1] "As we survey the course of history, it seems reasonably clear that all major social structures have in due course been cumulatively modified or abruptly terminated. In this event, they have not been eternally fixed and unyielding to change."

I am convinced that the industrial world is heading for another push of basic innovation (Chapter 7), and our children will call it the fifth (or so) industrial revolution or the First Something-Else Revolution. In any case, the transition will be guided by some tradition (Chapter 8) and yet truly revolutionary.

REFERENCE

1. R.K. Merton, "Manifest and Latent Functions," in: N.J. Demerath and R.A. Peterson, eds., *System, Change, and Conflict*, New York: The Free Press, 1967: 9–75.

Chapter 7

A Bold Projection into the Future

"Panoptical purview of political progress and the future presentation of the past"
(James Joyce, *Finnegan's Wake*)

Given that the current stagnation phase in which the industrial world has found itself for a number of years is a developmental interlude and that its timing is entirely in line with the long-term calendar of events depicted by our metamorphosis model (Chapter 2); and given that the evolutionary interplay between stagnation and innovation is characterized by considerable discontinuities in the rate of basic innovations, so that ebbs in the flow of basic innovations result in a slowdown of economic growth, recession, and even depression at worst, and so that these distresses in turn result in a flood of basic innovations (Chapter 4); and given the fact that innovations research until now has not succeeded in detecting any political process, market mechanism, or mixed economy that promises to alter this rhythmical pattern of inertia and ebb and flood in the stream of basic innovation; we may proceed from these givens to the next question. Will there be another flood of basic innovations in the future, and if so, when? In this chapter, I try to provide for a projection of past and current trends into the future. Such a projection is, of course, neither a prognosis of the innovation cycle nor an unconditional scenario. Rather, such a projection depicts only one of several possible configurations of future innovative change. We always depend on such projections when faced with the need for contingency planning in family affairs, business matters, or policymaking. The following projection is conditional upon the assumption that the rhythmical pattern in the interplay of stagnation and innovation will continue to evolve in the next 20 years as it has done in the last 300 years or so.

With the exception of the capacity to bring nuclear catastrophy and similar shocks into the world, I see no man-operated organization yet fully at work that might modify the pace of basic change in a few years. Thus, the following projection might be a good approximation. On the other hand, the basic challenge of our epoch clearly is gaining better understanding of the pacemaking of change, as this is the key to civilization's problems of economic stability and selection of the best chances for basic innovation.

HURDLES IN THE TRANSFER OF KNOWLEDGE

As the tempo of innovative change is determined by the speed of advancement of invention, research, development, investment planning, implementation, production, distribution, and so on, of the new technologies, an analysis of the average tempo of change will have to reckon with differences in the tempo of change at different stages of innovation activity. In modern times, innovation processes have to pass through an extended network of personal and institutional relationships, and a specific concept may fail to overcome any of several hurdles in the transfer of knowledge.

In older times, this was not necessarily so. During the Renaissance, when production was still organized on the basis of craftmen's workshops whose dimensions were limited by size, by feudal divisions of territory, and the jealous attitude of guild members toward domestic markets, inventive activity concentrated on implements of weapon systems (recall Leonardo's designs), waterworks, churches, ships, and bridges, and it specialized in making new devices and tools as shown in Polydor's famous woodcut (see Figure 7–1). In these situations, the inventor and innovator were generally the same person. The master or apprentice who had an idea for a new design or a new tool like those shown in Figure 7–1 produced and distributed it alone. Modern scientific technology requires teamwork to develop new products and processes. Trial and error still obtains in modern research and development, but the necessary fundamental knowledge can no longer be acquired through self-teaching; a formal course of study is usually essential. Moreover, the process of converting this theoretical knowledge is highly complex. The transfer of knowledge from theory to practice involves surmounting a number of obstacles. We can consider them as steps toward the practicality of ideas and knowledge.

Figure 7–1. Typical Inventions at the Early Age of the Patent System.

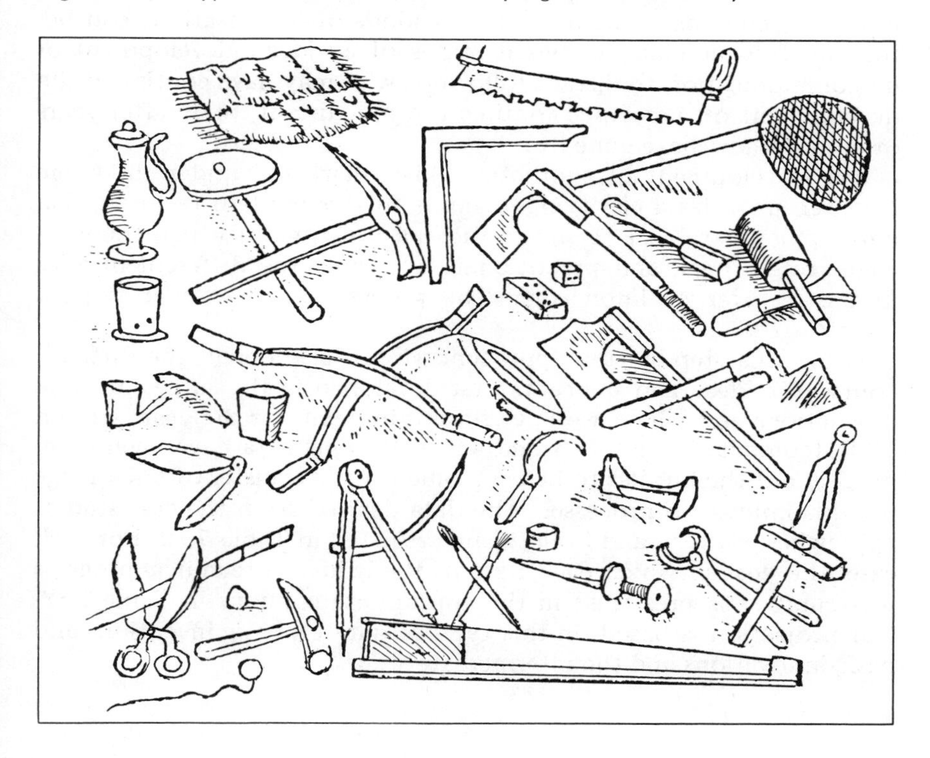

Every stage requires more certain and precise proof of the practicability of the idea and more resources:

Stage 1. Discovery or development of a new theory (perception)
2. Observation of a possible practical application (discovery, basic invention)
3. Proof of the technical practicability of the idea (feasibility)
4. Beginning of market-oriented experiments with the idea (development)
5. Decision to implement the new technology (decision)
6. Production begins with the new process or the new type of product being introduced on the market (basic innovation)

Using this step-by-step outline we can trace the history of basic innovations and establish what works and experiments occurred in conceptual, research, and developmental areas and delineate the successive stages in the complex process of knowledge transfer. In doing this, we find that many people participated in the development of modern technologies; many different kinds of organizations had bethis involved at many different stages of an idea's development or its communication to distant followers, who may have continued the development on a distance continent or in a distant time—after years or even decades have gone by.

The development of synthetic rubber marketed under the name Neopren provides a clarifying example of this modern technological development. Table 7–1 shows the span of time in which theoreticians, researchers, and practitioners overcame the different hurdles in the transfer of knowledge. The process extended over at least twenty-six years.

Using this step-by-step approach, we can also depict the difficult course for ideas and concepts that developed in the surges of basic innovations and which we have set out in detail for the years around 1935 (compare Tables 4–4 and 4–7). Various data have been compiled about the hurdles to be overcome for the transfer of knowledge for this innovative processes. The data derived from the case studies by Jewkes, Sawers, and Stillermann are listed in Table 7–2. For each case, the year in which each rung of the ladder of technical progress was climbed is presented in the appropriate columns in Table 7–2. Our projection is based on this type of data on basic inventions and basic innovations and the intermediate steps.

Table 7–1. Events in Neopren Development.

Year	*Stage*	*Events in Neopren Development*
1906	2	Julius A. Nieuwland observed the acetylen reaction in alkali medium and worked for more than ten years at the problem of higher yield of the reaction (basic invention)
1921	3	Nieuwland demonstrates that his material, "divinylacetylen," a polymer, can be produced through a catalytic reaction (feasibility)
1925	4	Dr. E.K. Bolton of du Pont listens to a lecture of Nieuwland at the American Chemical Society; du Pont assumes the further development of this type of rubber material (development)
1932	6	E.I. du Pont de Nemours and Company introduces Neopren, a synthetic rubber, into the market as a new, commercial product (basic innovation)

Table 7–2. Sequence of Events over Time in the Transfer of Knowledge.

Basic Innovation	*2*	*3*	*4*	*6*
Automatic drive	1904	1931	1932	1939
Hydraulic clutch	1904	1926	1934	1937
Rollpoint pen	1888			1938
Catalytic cracking of petroleum	1915		1930	1935
Watertight cellophane	1900	1912	1925	1926
Cinerama	1937			1953
Continuous steelcasting	1927	1935		1948
Continuous hot strip rolling	1892	1920	1921	1923
Cotton picker (Campbell)	1920		1924	1942
Cotton Picker (Rust)	1924	1930	1933	1941
Wrinkle-free fabrics	1906	1926	1929	1932
Diesel locomotive	1895	1903	1925	1934
Fluorescent lighting	1852		1926	1934
Helicopter	1904	1922	1926	1936
Insulin	1889		1920	1922
Jet engine	1928	1934	1939	1941
Kodachrome	1910	1923	1925	1935
Magnetic taperecording	1898		1937	1937
Plexiglass	1877	1929	1931	1935
Neoprene	1906	1921	1925	1932
Nylon, perlon	1927	1930	1930	1938
Penicillin	1922	1928		1941
Polyethylene	1933	1935	1935	1956
Power steering	1900	1925		1930
Radar	1887	1922	1933	1934
Radio	1887	1900	1907	1922
Rockets	1903	1924	1929	1935
Silicones	1904		1935	1946
Streptomycin	1921	1925	1939	1944
Sulzer loom	1928	1931	1936	1945
Synthetic detergents	1886	1913	1917	1928
Synthetic light polarizer	1857			1932
Television	1907	1919	1923	1936
"Terylene" polyester Fiber	1941			1955

Table 7–2. continued

Basic Innovation	*2*	*3*	*4*	*6*
No-knock gasoline	1912		1919	1935
Titanium	1885	1919	1930	1937
Transistor	1940		1948	1950
Tungsten carbide	1900		1918	1926
Xerography	1934	1940	1946	1950
Zipper	1891	1913		1923

Legend: 2 basic invention; 3 feasibility; 4 start of commercial development; 6 basic innovation.

TREND EXTRAPOLATION: LOOKING AT THE NEXT SURGE OF BASIC INNOVATIONS

Radical changes in trends cannot be predicted as easily as one can extend a straight line x units into the future. Under the stipulation that relationships between interrelated trends remain as they currently are, trend extrapolation can nevertheless result in a useful projection. Any attempt to extrapolate nonlinear trends is fruitless unless one can follow several interrelated trends over time and establish that their proportional relationships remain the same. This we shall do.

To extrapolate nonlinear trends we must express them in percentages so that their proportionality can be assessed. We have accomplished this with the time sequences from Table 7–2, which have been converted into percentages in Table 7–3. We need the parameters of the percent distributions—the mean value (μ) and the standard deviation (σ). These values appear in Figure 7–2. The mean value is the year in which half of the stages in the time sequence have already occurred, and the other half will follow in the years to come. The standard deviation is the number of years before and after the mean value year in which approximately 34 percent of the innovations have occurred. For example, the distribution $P(\mu, \sigma) = P$ (1935, 8) in Figure 7–2 signifies the following. In 1935 exactly one-half of the basic innovations in the surge of innovations had occurred, and in the years 1935 ± 8 approximately two-thirds of the innovations from the period 1900–1955 took place.

By converting all of the data that we have gathered from case studies concerning basic innovations (Chapter 4) into percentages, we arrive at the distributions in Table 7–4. Assuming that the rhythmical pattern of change described by these parameters will not deviate

Table 7–3. Time Series (simple, accumulated, and in percent) of Events in the Transfer of Knowledge.

	Stage 2 Time Series			*Stage 3 Time Series*			*Stage 4 Time Series*			*Stage 6 Time Series*		
Periods	*Simple*	*Accumulated*	*in %*	*Simple*	*Accumulated*	*in %*	*Simple*	*Accumulated*	*in %*	*Simple*	*Accumulated*	*in %*
1850-1854	1	1	2, 4									
1855-1859	1	2	4, 9									
1860-1864	0	2	4, 9							Compare Table 4-7		
1865-1869	0	2	4, 9	1	1	3						
1870-1874	0	2	4, 9	0	1	3						
1875-1879	1	3	7, 3	0	1	3						
1880-1884	0	3	7, 3	0	1	3						
1885-1889	6	9	22, 0	0	1	3						
1890-1894	3	12	29, 3	0	1	3						
1895-1899	1	13	31, 7	0	1	3						
1900-1904	8	21	51, 2	2	3	11	1	1	3			
1905-1909	3	24	58, 5	0	3	11	1	2	6	1	1	2, 4
1910-1914	2	26	63, 4	3	6	21	0	2	6	0	1	2, 4
1915-1919	1	27	65, 9	2	8	29	3	5	15	0	1	2, 4
1920-1924	4	31	75, 6	5	13	47	4	9	27	3	4	9, 8
1925-1929	4	35	85, 4	6	19	68	8	17	51	3	7	17, 1
1930-1934	3	38	92, 7	6	25	79	8	25	75	7	14	34, 1
1935-1939	1	39	95, 1	2	27	92	6	31	93	13	27	65, 9
1940-1944	2	41	100	1	28	100	0	31	93	5	32	80, 5
1945-1949							2	33	100	4	36	87, 8
1950-1954										4	40	97, 6
1955-1959										1	41	100

Figure 7–2. Depiction of Parameters (μ, σ) of Time Series of Events in the Transfer of Knowledge.

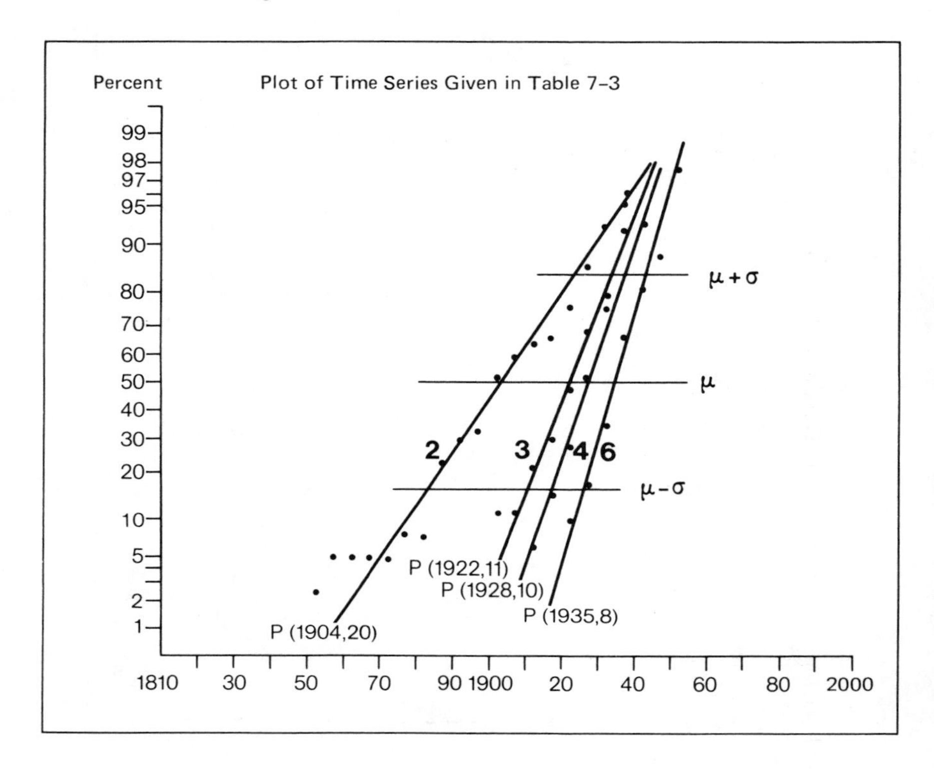

Table 7–4. Frequency Distributions of Events in the Transfer of Knowledge.

μ = *year in which one half of all events had taken place*
σ = *standard deviation in years*

Hurdle in Transfer of Knowledge		*Empirical Distribution (μ/σ)*		*Estimated Distribution (μ/σ)*
Stage 1	(1775/41)	(1825/39)		(1935/32)
Stage 2			(1904/20)	(1959/17)
Stage 3		(1826/16)	(1922/11)	(1971/10)
Stage 4			(1928/10)	(1979/7)
Stage 5				(1985/5)
Stage 6	(1825/18)	(1881/11)	(1935/8)	(1989/5)

significantly in the future, we can use the parameters from these distributions to derive a trend extrapolation. (Figure 7–3 and 7–4).

Examined horizontally, Figure 7–4 illustrates how the totality of major scientific-technical achievements has fluctuated over time. If trends continue in the same fashion, a new surge of innovations is indicated. Viewed vertically, that is, following an idea from theory to practice, the graph reveals how the individual hurdles in the transfer of knowledge become more formidable—that is, they become higher and steeper, the closer they are to actual practice. It is a process that unfolds (as does every evolutionary process that follows the laws of entropy). Following the years of technological stalemate and lack of basic innovations, we must expect a new, powerful surge of innovations if the economic depression continues along the current path.

Figure 7–4 illustrates how the distribution function describing the expected surge of innovations P (1989, 5) was inferred from the positions of individual curves and their deviations both over time and during the stages of knowledge transfer. This distribution harmonizes best with the different nonlinear curves so that the overall pattern that differentiation and distribution produced in the past remains consistent when extrapolated into the future. If the present tendency continues, that is, delays in the transfer of knowledge followed by acceleration because of the pressure from the damned-up demand for technological basic innovations, the distribution depicted in Figure 7–4, P (1989, 5), will then suggest the following consequences:

1. Only a small number of the basic innovations that will be implemented in Western economies by the year 2000 will be implemented during the 1970s. Therefore, if conditions continue as they have, during this decade it will not be possible to break out of the stagnation trend that brought recession and depression with it.
2. Approximately two-thirds of the technological basic innovations that will be produced in the second half of the twentieth century will occur in the decade around 1989.

The surge of innovations will begin in earnest after the year $\mu - \sigma = 1989 - 5 = 1984$. This is a fascinating date when one considers the historical fact that in the past surges of innovations have always occurred during years of iron rule in national government (by 1825, 1886, and 1935). One should also consider this tendency in light of Orwell's vision of 1984 and Andrey Amalrik's book title question as to whether the Soviet Union can survive the year 1984 and overcome its autocratic rigidity.

Figure 7–3. Parameters of Drifting Time Series of Events in the Transfer of Knowledge.

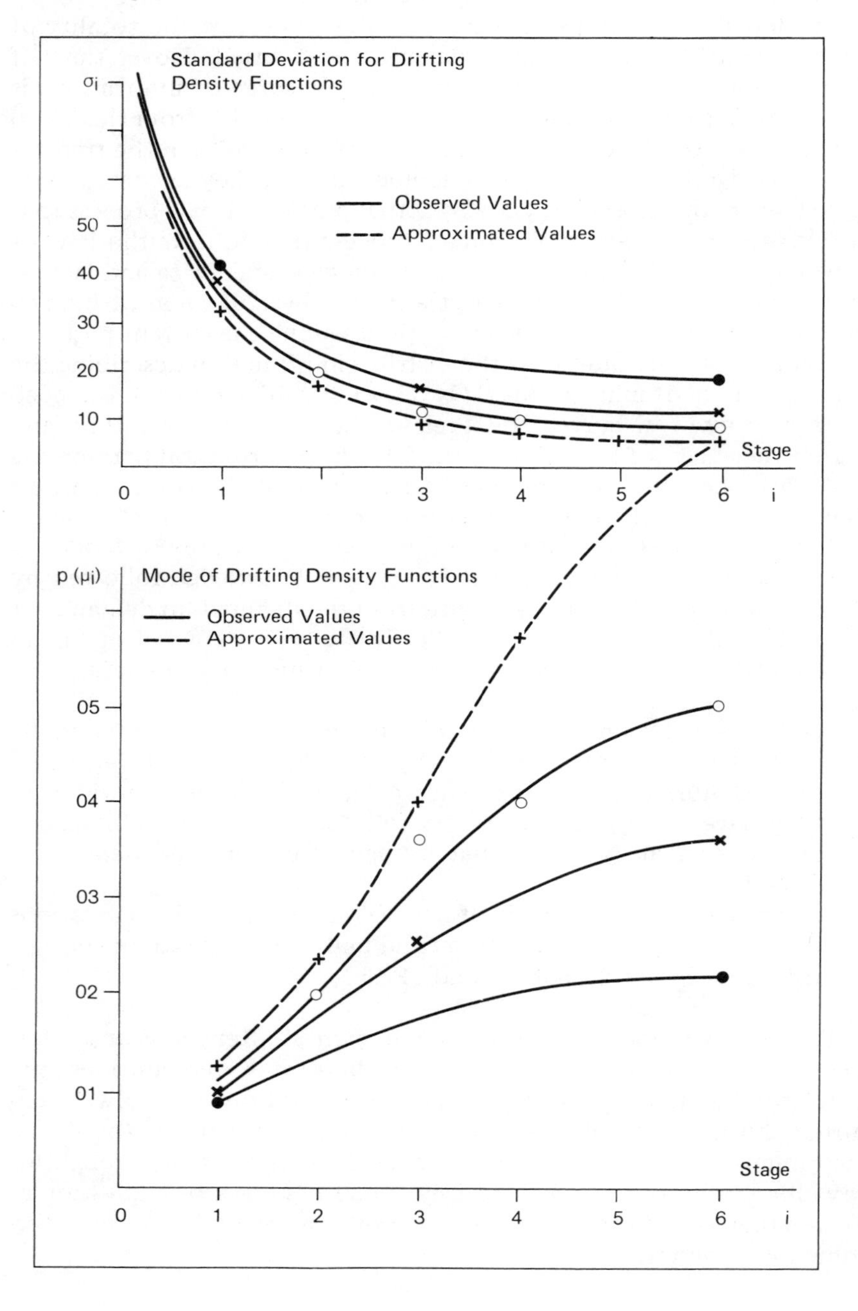

Figure 7–4. Composite View of the Frequencies of Events in the Transfer of Knowledge.

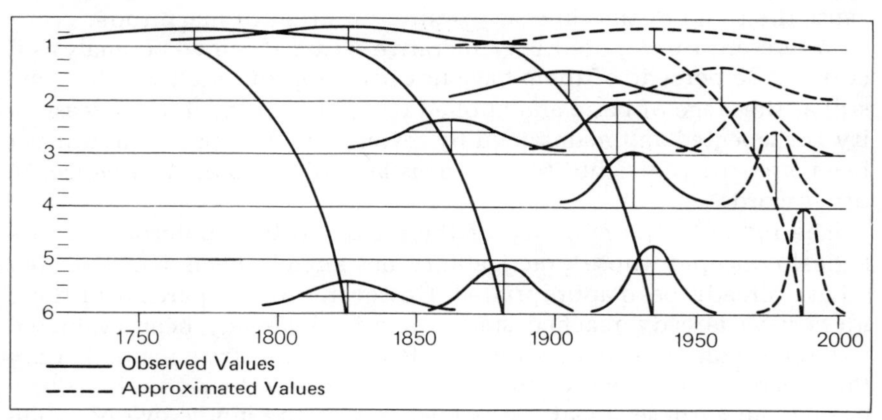

3. From the distribution of events at stage 3, *P* (1971, 10) we can infer that for most of the technologies that will be launched during the rest of this century it is already clear today that there are practicable innovation concepts available. Although most of the innovations are still in the testing stage, some of the new concepts that have practical application still need to be substantiated (feasibility).
4. From the distribution of events at stage 4, the beginning of the commercial development stage *P* (1979, 7), we can infer that today just about 50 percent of the possibilities for technological basic innovations have reached the beginning of commercial development. Moreover, by 1980 only barely half of the projects for basic innovations that will happen during the next 25 to 30 years will not be worked upon by some private firm or public institution who mean to put it into practice.

These projections, which are naturally only conditional, mark the boundaries for entrepreneurial opportunities. To illustrate what we mean by this, let us pose a rough-guess question. How would a company be situated in the year 2000 if during the next few years it acquired the rights to use one of the ideas now being evaluated at stage 3 and during the next few years it succeeded in developing and marketing this new concept? The company would have the status, both financially and in prestige of IBM, ITT, GE, Xerox, Boeing,

Polaroid, Dupont, 3M, or Ford. However, the company would be approximately two or three times the size of these companies because the trend of the times is toward that order of magnitude.

Where are these potentially lucrative ideas? If our trend analysis is correct, 25 percent of them have not been appropriated yet; they are still at the stage of basic and applied research. Although their feasibility is suspected and maintained by researchers, because it has not yet been proven, the rights to the ideas are still unsold. *A bonanza*, in other words.

The rights to the majority of the ideas—to the numerous technological concepts whose practicability has already been demonstrated —have already been appropriated. However, only 50 percent of these ideas have already reached stage 4 and have already been evaluated for their commercial possibilities. Because industrial research only takes care of 5 percent of the basic research and the research at stage 3, one can surmise that these 50 percent of the innovative possibilities that have been appropriated are less likely to be held by capital-rich large companies than by government research institutions, universities, and so on, and thus many of the ideas are available for sale, for joint ventures, and for technology transfer at zero cost.

Some odd 30 percent of the projects that have already passed stage 3 but have not yet reached stage 4 in their economic development are either owned by large companies that do not understand the economic potential of these projects or by individuals who do not have the resources required to develop their patented ideas commercially. These individuals are a valuable target group for enterprising people who—as entrepreneurs always do—try to get the most valuable production factor at minimal expense. Unless these researchers and inventors (whose talent generally lies in technical areas) quickly develop a superior business acumen, they will be taken advantage of or even ousted from participation. Thus, they will be added to the numbers of people who have not been compensated for their contributions to the innovation process and who failed to enjoy any share of the enormous profits from their original invention.

Most of our innovations research is devoted to the sources of these new opportunities, the demand for it, and to the microeconomic conditions by which these chances can be made a reality. In the following two chapters we shall present some macroeconomical generalizations and their societal impacts.

SUMMARY

Under the condition that the rhythmic pattern in stagnation and innovation does not suddenly change its pace (a possibility for which we have found little evidence as yet), we can provide a look at the next surge of basic innovations.

Only a small number of basic innovations will be achieved prior to 1984. Approximately two-thirds of all the technological basic innovations that will be achieved in the second half of the twentieth century (half this period having already elapsed) will occur in the decade around 1989.

Many of these great business opportunities are already appropriated to some company in some country; a good many of these chances are still available at reasonable cost, and they should challenge the entrepreneurial spirits in private and public enterprise.

Chapter 8

Tradition and Transition

"Ideas are, in truth, forces" (Henry James)

If my interpretation of the present conditions as a stalemate in technology is correct and if my inference of a forthcoming breakthrough of a large number of basic innovations justifies at least some further investigations of this contingency, the next question must then turn to the sources and the characteristics of the coming socioeconomic transition.

I am convinced that the industrial world is heading toward some form of a "leisure economy." Both the degree of technical advance and automatization and the perceptible changes in people's ways of allocating their time indicate that a decreasing amount of working time will be devoted to production in now established industries and service sectors. This implies that an increasing amount of time will be allocated either to new types of industry and service delivery systems or to more or less productive pastimes. Therefore, two alternative scenarios for the forthcoming transition suggest themselves to be considered separately. They are not mutually exclusive; they only imply that the forthcoming cluster of basic innovations is composed differently.

At one extreme, most of the basic innovations will be technological, and the coming transition is "hard" in the sense that it advances the Western civilization toward hyperindustrialization and a way of life that reminds me of Huxley's *Brave New World.* At the other extreme, most of the basic innovations will be nontechnical, and the coming transition is "soft" in the sense that it advances the West toward a postindustrial society and a way of life for which the label "participative economy" may serve as a first approximation.

TRADITION GUIDES THE TRANSITION

As soon as we envisage even a slight possibility that the forthcoming transition is not predetermined but could conceivably proceed along alternative lines of development, we emancipate from the blindfolding pragmatism ("What will be, will be"), into which we are double-bound by the inacceptability of social Darwinism ("Whatever is, is right"; Alexander Pope), and its polar opposite, developmental anarchy ("Whatever is not, is better"; Bakunin). Experience with any attempt to emancipate from pragmatism warns us, however, against utopian dreams about the governability of change.

As we shall demonstrate in the following section, all of the economic transitions observed in the past have been preprogrammed by a powerful tradition. The order of change is distinctly conservative. From this I infer that any major direction of innovative change in the foreseeable future will be in line with some highly influential school of thought that is already or is just about to become a widely accepted tradition.

Any time in history when the economy became ripe for restructuring, the evolutionary interplay of stagnation and innovation pivoted on the existence and the availability of powerful knowledge. "Ideas are, in truth, forces," said Henry James; but it was already in 1832, when scientific knowledge had visibly become a decisive factor of production, wealth generation, and political power, that Charles Babbage, in his essay," On the Economy of Machinery and Manufacture," made the power of knowledge bluntly clear to the established powers and to the reformers and innovators alike. It has been reiterated millions of times that the experience of the past has stamped with an indelible character of truth the maxim that, knowledge is power, Babbage said.[1]

Knowledge has become an instrument variable to all parties involved in the battle for power and wealth. The same knowledge is used both in the name of continuity, stability, and the preservation of established positions of power and in the name of discontinuity, progress, and the redistribution of wealth and power. Therefore, it should be delightful to discover the different ways by which this knowledge is being employed in the conflicts of wills and forces. Knowledge is being used both in the pursuit of progress and for progress halted in the name of Progress.

Sometimes knowledge flows stronger into the stabilization of established industries and the defense of market power and wealth and prestige derived from it until the marginal productivity of knowledge employed for these ends diminishes (stagnation). At other times knowledge flows stronger into structural transition and into areas where the productivity of the knowledge is high (basic innovation). Yet at all times the most powerful and productive knowledge is knowledge that is propagated by some highly esteemed fraternity of opinion leaders who either pursue their group interests (as all deans of the Invisible Colleges in Science do) or the vested interests of some group with which they strongly identify. In the past, the Invisible Colleges in traditional science guided the innovative activities of the Invisible Hand in the economy. Knowledge colonialized practice.

THE SENIORITY PRINCIPLE OF INNOVATIVE CHANGE

Tradition guides the transition. This clearly comes out from further analysis of the data of basic invention and basic innovation presented in Chapter 4. If we compare the sequence of basic inventions with the sequence of basic innovations, we find that despite several decades of knowledge transfer from invention to innovation, the precedence relationship is amazingly unaltered over time. The first basic innovations in a swarm also happen to have been the first corresponding basic inventions in the sequence. For obvious reasons, I call this the seniority principle in innovative change.

The seniority principle always means first come, first serve. In innovative change it is supposed to denote the observation that the first basic inventions have such a strong normative quality (paradigmatic property) that they predetermine a whole chain of further events. For one thing, they chart the epistemological territory for other basic inventions that will eventually be found by subsequent research and development, and for another thing, they prearrange through their intrinsic quality the sequence in which this set of basic inventions can reasonably be put to use as basic innovations. It is the obvious adherence of the practical world to such paradigms that allows us to state that tradition guides the transition.

Table 8–1 depicts and evaluates the seniority principle for the basic innovations that produced the Industrial Revolution (Brno study). With only three unorderly exceptions out of thirteen cases, the seniority principle holds with 95 percent reliability. This evolutionary principle (progress governed by the most conservative aspects of change) is too paradoxical to be easily believed. It has far-reaching implications.

At one end, it would tell economic policymakers that a pragmatic innovation strategy of "try anything, and do it quickly" is bound to

Table 8–1. The Seniority Principle in Innovative Change during the Industrial Revolution (Brno).

Ordered Sequence of Inventions	*Ordered Sequence of Innovations*	*Unorderly Exceptions*
1	1	
9	9	
10	2	2
7	10	
12	7	
2	5	
5	3	
3	12	12
6	4	4
4	6	
14	14	
11	11	
13	13	

Legend: Ordering the data from Table 4-5 into the sequences in which they emerged in time (case 8 of Table 4-5 is omitted for obviously not belonging to the set)

Analysis: If x is the number of exceptions to the rule that two ordered sequences of 13 pairs of events are identical, then $P(x/13)$ is the probability of error if one judges two similar sequences as being basically identical:

x =	1	2	3	4	5	6
$P(x/13)$ =	0.002	0.011	0.046	0.133	0.291	0.500

Result: As there are three unorderly exceptions only, the Seniority Principle is significant on the 95% level (0, 046)

fail in reality.[a] For example, observe the following thirteen figures on the average speed of innovative change taken from Table 4–5 and ordered according to the invention sequence in Table 8–1:

1.11; 1.25; 1.13; 1.21; 1.08; (senior cases)
1.54; 1.47; 1.63; 1.57;
2.04; 2.38; 2.63; 2.63. (junior cases)

The speedy emergence of the most junior basic innovations happened (with high-speed indicators between 2 and 2.7) only after the most

[a] Yes, I am referring to the pragmatist (New Deal) approach as Franklin D. Roosevelt described it in 1932: "The country needs . . . bold, persistent experimentation. It is common sense to take a method and try it: if it fails, admit it frankly and try another. *But above all, try something.*"

senior basic innovations had occurred (with low speeds between 1 and 1.25) and the intermittent changes had taken place (with medium tempo around 1.55). The obvious danger of the earlier technocratic innovation strategy is pushing wrong projects in which the loss consists not only in the cost sunk into the unsuccessful ventures but also in the so-called opportunity cost; that is, the disbenefits resulting from the foregone alternatives that have been pushed aside.

The topic of pushing wrong projects is old, and the debate about the New Deal had its predecessors. For example, Daniel Defoe, in his "Essay upon Projects" in 1697, which he was said to have written under some sort of government contract, suggests an innovation strategy as a remedy for economic stagnation and does not fail to deliver the appropriate warnings. The New Deal in the words of Defoe:

> Projects of the nature I treat about, are doubtless in general of publick advantage, as they tend to improvement in trade, and employment of the poor, and the circulation and increase of the publick stock of the Kingdom.[2]

The stagnation-innovation theory is also well known to Defoe:

> Much greater were the number of those who felt a sensible ebb of their fortunes, and with difficulty bore up under the loss of great part of their estates. These, prompted by necessity, rack their wits for new contrivances, new inventions, new trades, stocks, projects, any thing to retrieve the desperate credit of their fortunes.[3]

And here is how the warning against wrong projects is described by Defoe:

> There are, and that too many, fair pretences of fine discoveries, new inventions, engines, and know not what, which being advanc'd in notion, and talk'd up to great things to be perform'd when such and such sums of money shall be advanc'd, and such and such engines are made, have rais'd the fancies of credulous people to such height, that meerly on the shadow of expectation they have form'd companies, chose committees, appointed officers shares, and books, rais'd great stocks . . . so I have seen shares in joint-stocks, patents, engines, and undertakings, blown up by the air of great words . . . and at last dwindle away.[4]

Wrong selection of basic innovation projects by government science and technology policy, therefore, means that the technology push strategy fails because the manufacturers are unable to cope with the production intricacies of, say, spin-off technology from research and development in nuclear or space technology. Manufacturers may be unprepared or unwilling to implement the innovation because, for example, of disbelief in their own abilities or capacities. Usually, the demonstration effect of a key innovation is required before several other innovations seem plausible. Indeed, as we shall see, basic innovation occurs in a natural order, and the sequence is nearly identical with the ordering in which the basic inventions had occurred many years earlier. Two millenia ago, Virgil said: *possunt quia posse videntur* ("they can because they think they can").

Wrong selection of basic innovation projects also means ignoring some aspects of quality. The success of any innovation necessitates that its product quality is valued by the potential buyers who have to adopt it. For radical innovations, this often necessitates that the adopters be prepared for this novelty by earlier events—events with a strong normative appeal. Paradigm changes and basic inventions in natural science sometimes had the property of opening the eye for vastly improved ways of work; paradigm changes and basic inventions in the humanities sometimes had the property of opening the eye for greatly inhanced ways of life. Only after such a perspective had become somewhat of a tradition, at least in some segment of the population, was the public at large mentally ready to adjust to the basic innovations to come.

The time lags between basic inventions and basic innovations as well as the lags between successive innovations depend on the normative power of factual knowledge about them. As long as practical knowledge has not been proved in practice, the least that is required for influence upon change of practice is that some eminent scholars who hold prestigious positions in the scientific estate openly say that this knowledge will work in reality if it were applied. "The product of man's making is characteristic of the epigoni who fall heir to a school of philosophy, not to the original thinkers who bring it into being," as Marnell observes.[5]

Because it takes one or two generations to build a new school of thought in philosophy and science, it takes that long for paradigmatic theory to become an influential tradition; that is, influential enough to guide the transition of the related body of knowledge into practical know-how in the economy. The same principle applies to economic policymakers and to the policies they are inclined to adopt. Keynes, for example, ended his General Theory with the following remarks:

> The ideas of economists and political philosophers, both when they are right and when they are wrong, are more powerful than is commonly

> understood. Indeed the world is ruled by little else. Practical men, who believe themselves to be quite exempt from any intellectual influences, are usually the slaves of some defunct economist. Madmen in authority, who hear voices in the air, are distilling their frenzy from some academic scribbler of a few years back. I am sure that the power of vested interests is vastly exaggerated compared with the gradual encroachment of ideas. Not, indeed, immediately, but after a certain interval; for in the field of economic and political philosophy there are not many who are influenced by new theories after they are twenty-five or thirty years of age, so that the ideas which civil servants and politicians and even agitators apply to current events are not likely to be the newest. But, soon or late, it is ideas, not vested interests, which are dangerous for good or evil.[6]

The dialectics of a particular perspective becoming a tradition and thus a value system for transition allows people to cope with inertia and innovation at the same time. It provides continuity in thinking at a time of discontinuity in doing. It preserves the order of things when things change. Therefore, the seniority principle of innovative change not only implies that the innovation strategy of "try anything, but do it quickly" is wrong, but also that the innovation strategy of "invest in understanding the order of things before investing in a thing" is better, if not best.

The appropriate philosophical paradigm is the principle of suspended judgment. It enables us to break out of vicious circles that haunt the issue of selection—vicious tautologies like the Darwinian principle of the survival of the fittest, which cannot be put to a test because the surviving unit was by definition the fittest among all alternatives; or tautologies like the pragmatist principle of Pope ("Whatever is, is right"), for it justifies the "Try anything" principle and excludes comparing it to the alternatives, which since they have not been tried did not come about. Defoe says:

> Endeavour bears a value more or less,
> just as 'tis recommended by success:
> The lucky Coxcomb ev'ry man will prize,
> and prosp'rous actions always pass for wise.[7]

On the basis of the principle of suspended judgment, social science is called upon to devote at least 5 percent of its scholarly research to serious inquiry into need assessment, whereas economic science is expected to deliver techniques for exploratory market research and results that are useful for need-oriented decisionmaking in investment in research, development, and innovation.

A long tradition in the line of thought exists that believes progress should serve people's needs in the first place and only in the second place should it serve to boost the ego of the inventor of a new technology or bring super-profits to the entrepreneur who implements the new technology. Could this relatively strong tradition not serve as the guiding paradigm for the next round of basic innovations that I think is surely coming in the foreseeable future (Chapter 7)?

If we examine the paradigm that guided the emergence of basic innovations of the Industrial Revolution around the 1760s (see Tables 4–5 and 8–1), then we see the mercantilistic idea at work in a number of policy reforms, institutional changes, infrastructural developments, and capital market and labor market adjustments that paved the way for industrial enterprise. In fact, they were the legal and institutional preconditions for the feasibility of the factory organization. In short, all these basic innovations were designed for giving the invisible hand (the market system) more freedom to operate effectively. In the sequence of events over time, we can trace the dissemination of the invisible hand concept from the early writings of the famous mercantilists in the 1650s via the idea's penetration into the proposals and experiments of administrative reformers and policy consultants around 1700, and finally to the concept's implementation in various basic innovations after 1750. With this understanding in mind, the chain of events in hindsight seems to follow a very natural order.

If we examine the seniority principle further (Table 8–2) and ask what guided the emergence of the technological basic innovations during the nineteenth and early twentieth centuries, we see the invisible hand play its role in the allocation of investment in invention and innovation. However, we see it guided by a new tradition—the concept that industry means conquering nature by a number of tricks and techniques that natural scientists are paid to discover and to develop. In turn, the role of physicists, chemists, and other scientists is to say what works and what does not. We can assess this paradigmatic influence of science upon new practice by again analyzing the data on basic invention and innovation given in Chapter 4. Table 8–2 summarizes our findings, and Figure 8–1 depicts the statistical testing procedure employed. We find again the seniority principle of innovative change at work, and the statistics reveal a highly significant influence of the early, paradigmatic events on the selection and timing of later events. Clearly, the new traditions in science had a profound impact on the way in which the economy had changed. Innovation in the past shows the footprints of scientific ambition much stronger than any mark of need orientation. And during the 20th century, basic innovations were mostly guided by advancements in science and technology which were directed at the needs of the nation states.

Figure 8–1. The Seniority Principle of Innovative Change. *(Rank Correlation of the Ordered Sequences of Basic Inventions and Basic Innovation, Table 4–4).*

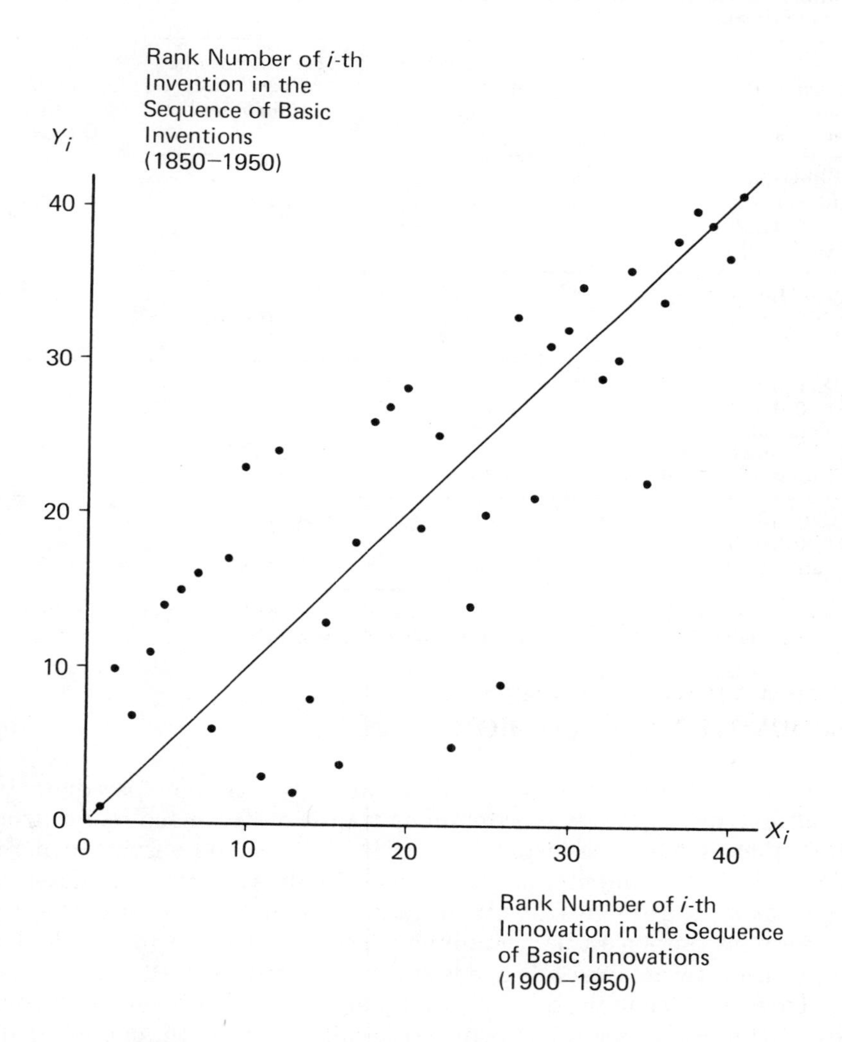

Result: Spearman Rank Correlation Coefficient $r = 0.8$ Level of Significance 0.999 (if all dots X_i, Y_i are on the diagonal, $r = 1$).

Table 8–2. The Seniority Principle of Innovative Change in the Nineteenth and Twentieth Centuries.

Method: The Spearman Rank Correlation Coefficient r indicates the degree of similarity of two sequences of events. If n events are in the sequences, the level of significance is $Q(r, n)$.

Cluster of Basic Innovations	*Number of Events in the Sequence (n)*	*Spearman Rank Correlation Coefficient (r)*	*Level of Significance Q (r, n)*
Industrial Revolution 1750–1800 (Table 4–5)	13	0.866	95%
Basic Innovations 1800–1850 (Table 4–1)	21	0.900	95%
Basic Innovations 1850–1900			
Chemical Industry	28	0.79	97.5%
Electrotechnical	22	0.88	98%
(Tables 4–2 and 4–3)			
Basic Innovations 1900–1950		see Figure 8–1	
(Table 4–4)	41	0.80	99%

Result: Very high, highly significant rank correlation coefficients indicate that basic inventions and basic innovations occurred in nearly identical sequences.

A NEW TRADITION: THE SCHUMPTERIAN PARADIGM

Therefore if the subsequent question is which new paradigmatic tradition will guide the forthcoming transition to an economic structure that is better adjusted to people's needs, the answer to me is clear. It is Schumpeterian economics, with a strong emphasis on innovative investment (the attribute that sets it apart from Keynesian economics among other economic traditions) and on need-oriented innovation (more people's needs and less governmental).

To several economists, for example, Joan and Dwight E. Robinson, Schumpeter's model might not qualify, as it is "though-minded" (Joan Robinson) in its major focus on the supply side, or even ignorant (Dwight E. Robinson) about the demand side of the economy. In any case, Schumpeterian economics lends itself to a supply-side oriented ("structural") policy after Keynesian economics has in the past decades served to convince economic policymakers that a ("process") policy of managing the level of aggregate demand is what the industrial economy needs.

It might be helpful to quote Joan Robinson at length because she sketches all the alternative economic paradigms that have sufficient tradition today to have an impact on policymaking:

> It is possible to defend our economic system on the ground that, patched up with Keynesian correctives, it is, as he puts it, the "best in sight." Or at any rate that it is not too bad, and change is painful. In short, that our system is the best system that we have got. Or it is possible to take the tough-minded line that Schumpeter derived from Marx. The system is cruel, unjust, turbulent, but it does deliver the goods, and, damn it all, it's the goods that you want.
>
> Or, conceding its defects, to defend it on political grounds—that democracy as we know it could not have grown up under any other system and cannot survive without it.
>
> What is not possible, at this time of day, is to defend it, in the neo-classical style, as a delicate self-regulating mechanism, that has only to be left to itself to produce the greatest satisfaction for all.[8]

It might also be useful to quote Dwight Robinson at length because he points precisely at the weakness in Schumpeter's concept.

> In attempting to construct his model of the dynamic economy, Professor Schumpeter omitted almost all reference to the consumer and his insecure taste. It was as if he chose to bring into the purview of economists only the first of the famous pair of extraneous forces which they had so long been happy to leave as *terra incognita*—namely, "The State of the Arts." "Changing Tastes," the other blade of the supply-demand scissors, he chose to ignore. For him the busy entrepreneurial chicken obviously preceded the egg of the desire to consume.[9]

There is no immediate need to defend Schumpeter against the insinuation that he saw a precedence relationship in what is a chicken-and-egg type of interrelation. The weakness of Schumpeter's work is that he focused mainly on change in the supply side. Nobody is perfect; he did not elaborate on the issues of a changing need structure. He probably knew that it would take another genius to fill in the gap, a genius who has not yet come and who will contribute an operational method for shift analyses of consumer's needs. As long as economic research lacks the capability to help entrepreneurs target innovations, consumers are likely not to get what they really need, whereas suppliers cannot do better than offer their products and services on a take-it or leave-it basis.

Discrepancies between what people need and what they can get result in sluggish demand, stagnant supply, and a waste of economic potentials. Need-oriented changes in the supply structure of the economy can bridge the gaps. A necessary condition for the potential supplier's ability to supply what is needed is the availability of need analysis. In marketing, we do our best. Unfortunately, industry often fails to fulfill the sufficient condition for need-oriented innovation because it is often more convenient or more profitable to hold innovation back. For many modern needs which evolve under the complexities of modern life, industry has no product and the state has no public good to offer; and people suffer from a dearth of "participative" goods and services.

This may or may not change in the future; the odds are, as we shall see in the next chapter, favorable that it will not. Then we are probably heading for a "hard" transition into hyperindustrialization where technological feasibility defines what will be offered. My impression from market research is that this line of development is not exactly what consumers want. Sluggish demand in many markets indicates that consumers prefer it the other way around. They want more need orientation in the economy and less subservience to technology.

Need orientation means that we no longer define our innovation targets primarily in terms of the conquest of nature, as Babbage, for example, expressed it in 1832:

> These productions of nature, numerous and varied as they are, may each, in some future day, become the basis of extensive manufactures, and give life, employment, and wealth, to millions of human beings. But the crude treasures perpetually exposed before our eyes, contain within them other and more valuable principles. All these, in their innumerable combinations, which ages of labour and research can never exhaust, may be destined to furnish, in perpetual succession, new sources of our wealth and of our happiness. Science and knowledge are subject, in their extension and increase, to laws quite opposite to those which regulate the material world . . . the further we advance from the origins of our knowledge, the larger it becomes, and the greater power it bestows upon its cultivators, to add new folds to its dominions.[10]

Need orientation means that we target the production of knowledge and its utilization more at the dominions of nature that we carry with us, as they are within us and exist between us. I hope that in this sense "The present is pregnant of the future" (Leibnitz).

SUMMARY

Tradition and transition, although often considered antagonistic needs, in fact are dialectic forces that interact. Transition in theory shapes new traditions in science, which in turn guide the transition of the economic structure, and economic forces influence the choice of areas where research is done and theory developed.

It is argued that if the political and economic forces make knowledge chiefly an instrument of inertia, it becomes decreasingly productive, and this leads to stagnation of political influence and economic prosperity. It is also argued that if knowledge is put into basic innovations, it has increasing marginal productivity, and this leads to economic growth and prosperity.

The main question is what kind of knowledge should be searched for, researched, developed, and what is a reasonable innovation strategy. From the data available, we deduce the existence of a seniority principle in innovative change (tradition guides the transition). It implies that technological basic innovations occur in a particular order, and this order is nearly identical with the order in which the basic inventions occurred many years before. It is a "natural order of change"—a protostructure for a whole network of chains of events.

Therefore, research policy should give a higher preference to research in the "natural order of change" because ignorance about it means that innovation policy is blind and its effectiveness is limited to the high-risk-low-return niveau. A strategy of "try anything, but quickly" produces too many flops, and in the hurry, it suppresses the often superior alternatives at the expense of "anything."

On the macroeconomic aggregate level, this innovation strategy, as it is the pragmatic choice of technocrates, seems to reinforce the drift toward a "hard" transition; hyperindustrialization is what it might result in.

The challenge of this epoch is to bring human needs into play. The invisible hand does not lead progress toward the maximally attainable quality of life. It leads it to "anything."

REFERENCES

1. Charles Babbage, *On the Economy of Machinery and Manufacture* (London: 1832), p. 316.
2. Daniel Defoe, *An Essay upon Projects* (London: 1697), pp. 10–11.
3. Op. cit., p. 12.
4. Op. cit., p. 13.

5. William H. Marnell, *Man-Made Morals* (New York: Doubleday, 1966), p. 378.

6. J.M. Keynes, *The General Theory of Employment, Interest, and Money* (New York: Harcourt, Brace & World, 1935), pp. 383–384.

7. Defoe, op. cit., p. 17.

8. J. Robinson, *Economic Philosophy* (New York: Doubleday, Anchor Books, 1964), p. 140.

9. D.E. Robinson, "The Importance of Fashion and Taste in Business History, *Business History Review*, 37 (1963), 5–36, p. 7.

10. Babbage, op. cit., p. 315.

❋ *Chapter 9*

The Direction of Change

"Whatever *now* is established, *once* was innovation" (Jeremy Bentham, A Fragment on Government, 1776)

"HARD" VS. "SOFT" TRANSITION

The preceding analysis yielded the following main results:

- The industrial economies have become structurally unstable; the capital market and many sectors have developed structural readiness for leapfrog type investment; which either means (in situations of overcapacity as, e.g., in areas like synthetic fibers, beer breweries) that some corporations will try to jump ahead of most competitors by quadrupling plant size for the sake of cutting per unit cost down to a half (*quantum jumps*); or which means that some entrepreneurs will take new technology (like the microprocessor) and implement it into a new type of commodity or service system (basic innovation) for which there is no direct competition (*quality jumps*).
- The Western economies enter a phase of transition where the propensity for quality jumps might produce a swarm of basic innovations; and we have every reason to assume that the selection of implements from the set of available possibilities for implementation will be guided by concepts of wants and needs that in this time and age have become somewhat of a tradition. All clusters of basic innovation in the past were timed and assembled by protostructural ideas that had evolved and had by then been taught to at least one generation of teachers and to two generations of students.

In retrospect, change appears to have been directed by an invisible hand that adhered to mental configurations that some cultural or scientific traditions had firmly established—like blueprints or user manuals—in the brains of decisionmakers. If on New Year's Eve in 1999 we look back upon the transition period that today in 1979 lies ahead of us, we shall find that the basic innovations now ahead of us seem to have been mapped out by cultural and scientific achievements dating back to grandfather's youth. "Man's manufactures are a product of his culture."[1]

The seniority principle in innovative change significantly determines "the assembly line process of history" (Karl Deutsch)[2] once the general direction of this process has been determined by the attraction of some dominant tradition of thought. As there are always several cultural traditions evolving *pari passu*, the general direction of innovative change is not determined in principle as long as none of the competing cultural traditions gains dominance either through its stronger appeal to the public at large or through the greater support it receives from an influential constituency whose power and wealth will be increased if this tradition guides the transition. Thus, the dice of change are rolling today, but at least in principle it is still open as to which specific numbers will show up in the future. The question, therefore, is (a) which numbers might show up and determine the general direction of innovative change and (b) whether the game is such that in actual fact there exists only one feasible direction of change and any number that will show up will be called to serve this predetermined end.

Hence, there are two scenarios for the main direction of change; neither is deterministic or undetermined, but one is less likely (the "soft" transition) than the other (the "hard" transition). One points at a postindustrial society where a greater proportion of nonmaterial wants and needs are being given a greater opportunity for self-articulation and self-organization ("soft"), whereas a smaller proportion of wants and needs is fixed in the "hard" way; that is, prearticulated and prearranged (by some form of state regulation, government agency, etc.) and prefabricated (by some supplier's organization). Both capitalism and socialism in their traditional, mainly "hard" way are largely alien to a "soft" transition toward more of a "participative economy" in which a larger share of the value-adding activities are nonhierarchically organized, mutually agreed upon, and voluntarily participated in as a leisure time occupation of emancipated people. By a "participative economy" I mean a microsocietal setting (a group, clan, clique, neighborhood, family, etc.) where people share in the use or creation of valuable things, services, pleasures, or burdens for which it is not customary or impractical to collect payment or where valuable contributions are being made by some people without expecting others to reciprocate now, later, or ever.

Whether the coming transition will have to be called "soft" or "hard" depends on the mixture of basic innovations that are forthcoming. I shall argue that a "hard" transition toward hyperindustrialization is more likely than a "soft" transition. This is because the rationalization bias will favor quantum jumps (and new, large-scale replacement technology) over quality jumps (new expansionary technologies and social innovation programs). The technology bias will also favor those quality jumps that can be brought about by implementing new technology rather than by nontechnological basic innovation; that is, specifically, if the latter type of change suffers from the comparative disadvantage of having no strong tradition of thought and knowledge behind it.

Note that, to illustrate the comparative advantage of technological innovations and the comparative disadvantage of nontechnical innovations, how easy it is to visualize the following examples of new replacement technology and new expansionary basic innovations. Also note the relative difficulty I am having in describing the need for social innovation programs in the "soft" area outside of the exchange economy, namely, in the "participative economy."

First, I shall give an actual example of *New Replacement Technology* and its use.

It has been estimated recently that the U.S. airlines will have to invest approximately 60 billion dollars in new equipment over the next twelve years. The old fleets have become decreasingly cost-effective in comparison to newer planes, and safety standards, noise controls, and regulation of fuel consumption will force the airlines to purchase improved equipment. United Airlines and Eastern Airlines have already placed gigantic orders with Boeing and the European Airbus consortium.

Only the wealthier airlines will be able to accumulate the necessary investment funds. Therefore, one must expect the airlines to go through a series of corporate mergers, for example, as the merger of National Airlines with Pan American World Airways already indicates.

Secondly, I shall give an actual example of *New Expansionary Technology* and its use.

The government of Saudi Arabia has recently decided on the ambitious plan to allocate about 30 billion dollars for the Fresh Water project in which nuclear-powered traction ships will pull gigantic icebergs from the Antarctic Ocean to the Red Sea. There, in the harbors of Jiddah and other artificial bays, the ice and the sunshine will deliver delicious water of which this country of barren mountains and naked deserts is in chronic short supply.

A whole cluster of basic innovations in ice technology, transportation, storage, and water distribution will be connected with the project. Although there is little competing industry and commerce that will become obsolete by the introduction of the new system, it will create enormous multiplier effects in the established and in novel branches of the economy.

Thirdly, I shall sketch the need for *Social Innovation Programs* in certain areas, but I am unable to suggest suitable problem solutions.

One preposterous result of the industrial civilization is that fewer and fewer parents carry the direct cost of reproduction (the expenses of child rearing, education, and other productivity-increasing investments in the following generation). At the same time, more and more nonparents harvest the economic benefits because the labor of the grown-up children will have to provide for the pension payments of parents and nonparents alike, for example. In addition, the market mechanism has so far failed to develop a reasonable way of renumerating the ten to sixteen hours of indispensable daily housework required to raise a family.

The "exchange economy" would collapse within twenty to thirty years if these contributions outside of the realm of mammon would cease to flow. Reasons of equitability will suffice to legitimate social innovations in this area of a "participative economy." And for the growing need for value-adding leisure activities, more social invention is needed strictly from the point of view of efficiency.

My point is that despite a growing need for social innovation even the most sophisticated members of our industrial civilization are slow in grasping such needs but fast in grasping the merits of the most grandiose industrial schemes. The Technological Bias has become a cultural bias; we can easily extrapolate our experiences with high technology systems; however, when we are faced with macro- or microsocietal problems and proposals for their solution, we have few experiences from which to extrapolate.

For technological inventions we are conditioned to first ask why they work. For social inventions we first ask why they may not. Life in the industrial economy has blindfolded us from seeing that the alternative to more industrialization (hyperindustrialization) consists in economizing the time (leisure time) that is spent in nonindustrialized activities (in the "participative economy"). A "soft" transition means that more social innovations that restore or renew the "participative economy" will be achieved than is the case in a "hard" transition toward hyperindustrialization.

Even the so-called mixed economy is basically foreign to a "participative economy" because it only means that the government controlled sector grows while the private sector stagnates. When more and more markets are transformed into state-operated pseudo-markets or quasi-markets, both private employers and government agencies collude in defining useful help in a neighbor's garden, repairing the neighbor's car or home, as a black market activity ("moonlighting"). The mixed economy system creates what is today called a "hidden economy," in which leisure time allocation to needed services is criminalized and the system fails to allow grey markets (like baby-sitting), indispensable labor (like child rearing), and charitable service to be structured so that a fair share of the value created can be reasonably appropriated to the donor by some value transfer mechanism. If the "leisure class" is limited to a few and if the increasing leisure time of the active population is prohibited from economizing, then the industrial economy will fight industrial stagnation with more industrialization, which is the "hard" transition.

There is no doubt that all industrialized countries in the East and the West show strong tendencies to evolve toward a form of a service economy and a larger leisure sector. In the industrial countries, productivity of labor has increased inversely proportional to the decrease of the labor week. For example, by 1975 the effective labor week shortened to about one-half of what it was in 1850:

Effective Labor Hours per Week *(Averages)*

	1850	*1890*	*1910*	*1940*	*1960*	*1975*
United States	72	60	54	40	40	39
West Germany	85	66	59	49	46	41

With technological progress, labor has become so productive that today one person employed in farming serves the demand for agricultural products as effectively as twelve to fourteen persons did a hundred years ago. While some proportion of the farming population moved to farm technology production, an even larger proportion became employed in different industries and in the service sector. This long-term trend clearly has suggestive power. It creates belief in a natural tendency of development into what some call the "service economy" and others call the "leisure economy." However, the long-term trend of the shortening labor week and the increasing share of services in gross national production easily camouflages an undesirable trend under a label that sounds comforting. In both the East and the West, government power and market power, as it is organized and as the incentives are, can enhance their transnational posture and their national standing by capital widening and deepening in directions of hyperindustrialization. "When all is said and done, pragmatism in American politics has taken its latest form in an attempt to buy Progress by monetary grants to individuals, communities, and states" (William H. Marnell), and there can be no illusion about what government agencies mostly get in return for their expenditures in research, development, and policy analysis. Most proposals they order and receive consist in feasibility studies on possible investments in capital-intensive, centrally organized, high technology projects that all add up to a hyperindustrial protostructure.

Clearly, massive investment in new, centrally organized service systems (Orwell's 1984) allows for an enormous extension of employment and of leisure time. Hyperindustrialization certainly is a direction of change for achieving both; however, it means a "hard" transition, and for the people this scenario means a different way of life from that implied by the concept of a postindustrial society. The "soft" transition, for which there still is some hope, suggests a transition into a postindustrial society in which a large share of what is valued by the people is produced and distributed in the "participative economy" as a leisure activity. The crucial difference between "hard" and "soft" is not the availability of less or more leisure time. There will be plenty of it in both ways. It is what people can do and cannot do with the available leisure time which makes all the difference between hyperindustrialization and a participative economy. The desirable feature of a participative economy is that it allows for a higher portion of "Eigenarbeit" (Christine von Weizsacker); of self-organized, value-adding activities and more self-determination.

My interpretation of the situation is that any "hard" transition will bring a type of Brave New World that, no matter if it was created by central planning or by market forces, will do away with democracy. That is, a "soft" transition is really the challenge and the chance of this stalemate in technology.

In the following two sections, I shall point out two mechanisms that make it more likely that we are facing a "hard" transition into hyperindustrialization and manipulation rather than changing "softly" into a form of postindustrial society characterized by a "participative economy" and emancipation.

WHY IS HYPERINDUSTRIALIZATION MORE LIKELY?

As the transition of the economic system is the result of basic innovations, there are two major forces that make me think that a "hard" transition is more likely than a "soft" one. These two transitional factors establish a technology bias (the patent system) and a rationalization bias (the price system). Both biases induce public and private investors to invest their available funds in the delivery system of a priceable commodity or service in which the technology, because it can be secured by proprietory rights, gives the owner some protection against direct competition and an instrument for securing a price and cost schedule. Technology usually gives at least partial control over input and output markets and over the firm's competitors. Ordinary organization, if it is not privileged by exclusivity rights or cartell type of arrangements, usually is more vulnerable and more subject to substitution.

This is why in an exchange economy, in which money is the means of exchange, Gresham's law (bad money drives good money from the market) also extends to other aspects of replacement:

- As in most industrial countries the cost of labor has been growing much faster than the productivity of labor, it usually pays to displace labor-intensive technology by capital-intensive technology (rationalization of production)
- As the utility of a product it may render to a user is measured by an average market price and where additional benefits to the user are not being transferred to the supplier, it usually pays to displace diversified production by mass production (standardization of products)

and good routine work drives better workmanship from the market (dequalification), and good average product quality displaces better qualities. This is the shadow price of industrialization.

Gresham's law also extends to the quality of services. Throughout the age of industrialization, the delivery of valuable services has either been commercialized and been made expensive and impersonal, or, if the commercialization of personal services was improper or impractical, the glorification of technological progress and industrial proficiency has put the stigma of backwardness, banality, and so on, on such activities as reading a poem to children, looking after one's disabled sister, or helping a neighbor renovate his or her roof.

As canned entertainment (which has a price) replaced home-made music and community festivities (which have no price), fewer people learned to make music. More and more people unlearned to celebrate and enjoy each other's contributions. Economic incentives and social conditioning enforced a hedonistic attitude, and two alternative attitudes fell prey to the preference for things that have a price attached to them. One victim is the high esteem for all those contributions that have no price attached to them, and the other victim is the motivation to organize life so that such contributions come forward easily. Another victim is the motivation for social invention and the benevolent attitude of give now and receive later or never (as opposed to give and get paid now).

Keynes once wrote: "There is no clear evidence from experience that the investment policy which is socially advantageous coincides with that which is most profitable."[3] At this point he was considering the bias of private enterprise in favor of quick profits, observed Joan Robinson, and continued, "There is a still more fundamental bias in our economy in favor of products and services for which it is easy to collect payment."[4]

This money orientation has had a profound effect on the motivational mix of most people, and although one sees these hedonistic tendencies reversed in many members of the younger generation, one does not see the same extent of recovery in people's ability to self-organize matters of practical concern and mutual benefit. The "hard" socioeconomic biases of rationalization in industry and industrialization of service delivery still exert a much stronger general influence on the allocation of resources than do the "soft" motivation for self-employment in business and the self-organization of exchangeable services and social contributions.

The Technological Bias

One major factor of industrialization and commercialization of many services by technological service delivery systems is the comparative advantage that patentable technological innovations enjoy over nonpatentable changes in business or social organization. Patents help the pioneer to secure the profits that flow from his or her invention, and social, nontechnical inventions usually do not qualify for patent protection. Therefore, ever since it came into existence 500 years ago, the patent system has had a profound influence on the socioeconomic evolution.

Many living beings today possess some kind of useless organ that once played a vital role in an earlier evolutionary stage, but which in the course of the development of the particular form of animal has gradually given over more of its function to other organs. Once an organ has lost its utility, it becomes a mere appendix. However, degenerated organs can fall prey to disease just as other organs do, and the disease can be fatal in fact.

The patent system has been in an analogous position to a vital organ during the early phase of industrial evolution. In the early phase of Western development during the Renaissance, this so-called organ provided a crucial metabolic process that produced the particular strengths necessary for the Western world's rise to its height during the industrial age. It stimulated new technology when there was little of it. However, what was of particular utility during the final stages of the Middle Ages is no longer helpful today to the same degree. Because of the degree of penetration of technology into our way of life, our civilization is suffering intensely from metabolic difficulties, that is, the technological bias in general, and specifically the excessive market power, which the patent system allows and induces to overexpand.

The patent system is an inheritance from the late Middle Ages. When one considers that it was first applied by Venetian dukes who were not scrupulous when pursuing their goals, one understands the tendency to cling to this system since its inception more than five centuries ago. The goal that the rulers of colonial Venice were pursuing with their "decreto sulla protezione della invenzioni" of March 19, 1474, was the encouragement of technical progress in directions that they desired—the invention of technical mechanisms, tools, and above all superior weapons.

As an epochal political innovation, it was introduced with this pithy text:

> MCCCCLXXIIII, the 19th day of March. There are in this city, and also there come temporarily by reason of its greatness and goodness, men from different places and most clever minds, capable of devising and inventing all manner of ingenious contrivances. And should it provided, that the works and contrivances invented by them, others having seen them could not make them and take their honour, men of such kind would exert their minds, invent and make things which would be of no small utility and benefit to our State. Therefore, decision will be passed that, by authority of this Council, each person who will make in this city any new and ingenious contrivance, not made heretofore in our dominion, as soon as it is reduced to perfection, so that it can be used and exercised, shall give notice of the same to the office of our Provisioners of Common. It being forbidden to any other in any territory and place of ours to make any other contrivance in the form and resemblance thereof, without the consent and licence of the author up to ten years. And, however, should any-

body make it, the aforesaid author and inventor will have the liberty to cite him before any office of this city, by which office the aforesaid who shall infringe be forced to pay him the sum of one hundred ducates and the contrivance be immediately destroyed. Being then in liberty of our Government at his will to take and use in his need any of said contrivances and instruments, with this condition, however, that no others than the authors shall exercise them.

favourable	116
contrary	10
uncertain	3

In 1550, a contemporary of the Venetian dukes, Polydor, made a woodcut (Figure 7–1) of "the invention of things" that he had heard about at that time. They are mostly small targets. However, it is easy to see the stimulating effect of the newly introduced patent protection in the bold ideas that appeared in the decade around 1500, particularly in the drawings of Leonardo da Vinci. Leonardo da Vinci's genius went beyond the painting of remarkable works of art to encompass the invention of powerful waterworks, mechanically powered engines, destructive weapons, and even designs for airplanes.

Ever since the Renaissance, technological inventions have been part of the outstanding achievements of enlightened people. The first U.S. patent was issued in 1790 to Samuel Hopkins and signed by George Washington. To many inventors, the rewards have been wealth and prestige. The patent system secured the inventor the economic proceeds from his or her idea, and society lauded the successful inventor with fame and honor. The prosperity and social prestige given to successful inventors attracted talent to pursue inventive paths. Research and invention held an increasingly prominent place in the dreams of the young and became a prime outlet for their urges like Columbus for adventure and discovery. Until today, when banal science fiction literature threatens to blunt the impact of this process, socialization in Western cultural circles was producing a growing number of scientists. The rate of patent applications in the European countries, Japan, and the United States has reached the level of 500,000 per year. In 1977, U.S. patent number 4,000,000 was issued to R. L. Mendenhall of Las Vegas for a new recycling method for asphalt materials.

However, the social conditioning of talent has been one-sided, with the patent system playing the questionable role of an amplifier. A patent can only protect an invention that is "new" and one can only establish newness when the invention possesses material or physical criteria, parts, or processes. Thus, there is no patent protection for nonphysical, nonmaterial inventions such as organizational reforms, new modes of behavior, new patterns of usage, new songs, new styles of painting or poetry, or new computer programs. One can patent new hardware but not new software, to use computer language. If successful, the search for technical innovations is worth the initial investment as well as a fortune. On the other hand, if anyone attempts to support possible inventions in organizational or service areas it is often necessary to make a large initial investment to totally occupy the entire market by leasing or franchises if one wants to secure this marketable nonmaterial innovation from competition. Without this massive initial investment, an innovation is easily copied, and the innovator often faces bankruptcy, poverty or hostility.

Of course there are exceptions to the rule, but the principal attraction (or the primary deterrent) offered by the patent system directs inventive talents into technology and keeps them from developing nontechnical inventions.

Naturally, there are also other reasons for the chronic technology bias of Western industrialization. However, several of these reasons are really side effects from the operation of the patent system. The major cause of the attraction that technology exerts is the system's discrimination against socially valuable but commercially marginal initiatives.

The protection that the patent gave to technical innovations not only produced a flood of new technical ideas but also caused an ebb in the creation of new organizational forms and services, because one could *not* patent nonmaterial, nontechnical inventions. *Since 1474, the economic evolution has been distorted by this technical bias*, which has not only produced intensively more destructive weapons, but has also led to unhappy trends in civilian areas. The relative contribution that the industrial sector makes to the West German national product is highest among Western countries, and the relative contribution made by service sectors unfortunately is the lowest. A similar distortion exists in Japan. While there is no other industrial country where the uses of economic resources are in such gross imbalance by rewarding certain kinds of behavior, the patent system universally seduces inventive minds into areas of technology where material technical progress can be made. Meanwhile, progress in humanitarian areas proceeds at a snail's pace: "Der Fortschritt ist eine Schnecke" (Günter Grass).

No one can tell whether technical progress would have undergone such a stormy development without the patent system. Perhaps today's world would be poorer by some machines, atom bombs, and pictures of the back side of the moon. Perhaps it would be correspondingly richer in other areas. Maybe without the patent system and technological monopoly, creative and energetic talents would have been more inclined to turn their inventive and dynamic minds to possibilities for organizational improvements in our working and daily lives. Humanity needs to catch up in these areas. There is plenty of room for a powerful surge of basic innovations in nontechnical activities.

One must therefore consider whether patent protection should be more restricted in the future to allow the fulfillment of this need to raise the quality of our lives or whether it should be extended to include nontechnical inventions. As it spreads a net of exclusive rights over the set of technologies and prevents the entry of competitors, the patent system plays an instrumental role in the imbalance between overinvestment in old and underinvestment in new types of business. The patent system not only enables these exaggerated trends and important omissions to occur but it even encourages their occurrence.

Macroeconomically, the patent system is no longer crucial to a high rate of technological invention, and under modern circumstances it even is partially counterproductive as it actually lowers the rate of innovation. Innovative projects compete with routine projects for investment capital that is devoted to the financing of business projects with maximum profit in mind. The decision is made on the basis of comparing profits and risks, and innovations always seem riskier than ordinary types of investment.

For this reason, projects for the research, development, and implementation of basic innovations are always considered last. Because they involve a high risk, costs are overestimated and profits underestimated. This is true despite the fact that successful basic innovations provide an opportunity for extremely high profits because they allow a monopolistic revenue to be derived from the technology. Thus, it is astonishing that during times of stagnation the private sector does not energetically marshall the knowledge available to produce technological innovations (the paradox of unused technologies) but instead continues to pour capital into the overgrown branches in which overcapacity has already begun to feel the pinch of underconsumption. This misdirection of labor and capital can be at least partially blamed on the patent system.

Under contemporary conditions, the patent system is *a hindrance to basic innovations.* New systems designs for high technologies can no longer be accomplished in a workshop or by a single individual; they require a highly differentiated research and development complex including numerous intermediate stages. In overcoming the obstacles of the knowledge transfer from invention to innovation, the front-runners hide the existence of unprotected ideas because they do not want to lose profitable rights to those coming after them in the process. Our modern differentiated innovative process is seriously hindered by these barriers to communication.

Under contemporary conditions, the patent system is also *a hindrance to improvement innovations.* Most of the markets for products based on fairly modern technology today are under oligopolistic control. A few large companies control the lion's share of the market segments, and because it is virtually impossible to dislodge these entrenched operations, they are able to lull quality competition into quiescence, thus giving themselves an unrestricted opportunity to raise prices. The control over the crucial technology is one precondition; the omission of feasible improvement innovations and the substitution of pseudo-innovations in their place is another consequence. At any rate, the patent system further amplifies the technological bias, and it drives the transition into the hard direction of hyperindustrialization.

The Rationalization Bias

The second main influence on innovative change is rationalization, namely, the concern for more efficiency in the utilization of labor, capital, and other inputs to production. To the extent that the rationalization of labor decreases the number of working hours per day, working days per year, and working years per capita, one might easily perceive of a natural, soft transition into a service and leisure economy. However, the following reasons make me believe in a higher probability of a hard transition into hyperindustrialization. I base my reasoning upon changes in the relative strength of the rationalization bias over time.

This means we will first have to look into its existence and its measurement, for in the discussion of "technological unemployment" some say it does not exist. "Innovations to save labor seldom take place to any considerable extent, except when there is a decided demand for them," wrote Thomas Robert Malthus shortly before 1820, the time when the steam engine was rapidly replacing labor in textile factories and other plants; he continued, "They are the natural products of improvement and civilization."[5] Certainly, "with the ratio of labor force to population rising in the long run less than 2 percent per decade,"[6] and "since man-hours per capita are found to decline 2 to 3 percent per decade"[7] during the last century, "the secular rise in per capita product"[8] was clearly a product of improvement and industrial civilization.

The industrial evolution produced, as we have seen in Chapter 4, an enormous flow of basic innovations in new technology and organization, and it induced last, but not least, an enormous increase of work intensity while working time decreased. Rereading the Malthus quotation above, one may then conclude that a rationalization bias does exist in general in the economy at large, but it does not always show up in special circumstances. On the long-term average, the enormous industrial growth during the process of industrialization and the increase in standards of living (free time, wealth, security, etc.) both resulted from an increase in efficiency within the economy at large. Macroeconomically, growth and productivity are inseparably linked to investments in rational production. Malthus knew it:

> When a machine is invented, which, by saving labor, will bring goods into the market at a much cheaper rate than before, the most usual effect is such an extension of the demand for the commodity, by its being brought within the power of a much greater number of purchasers, that the value of the whole mass of goods made by the new machinery greatly exceeds their former value; and notwithstanding the saving of labor, more hands, instead of fewer, are required in the manufacture.[9]

Given the obvious amalgamation of expansionary and rationalizing factors within the economy, how can one then measure the rationalization bias? The Ifo-Institute of Economic Research in Munich uses the following approach. Assuming that management can usually tell whether a specific investment project is directed chiefly at expanding a particular line of production or chiefly at cutting the per unit cost of products, the Ifo-Institute sends questionnaires to all German industrial corporations. By now it has collected data on the volume of investments directed at expansion and at rationalization for the twenty-two years from 1956 to 1977 (see Table 9–1).

Table 9–1 reveals an increasing, nearly doubling flow of rationalizing investments between 1956 and 1963 and a high level of continuous rationalization during the last fifteen years. It reveals a fluctuating volume of expansionary investments that obviously reflects the business cycle. The pattern of the changing investment structure is what concerns us here. The motive of efficiency and rationalization seems stronger and more stable than the motive of growth per se or of growth of capacity for the schedule of demand. At any rate, since 1960 the motive of rationalization seems to have dominated the motive of expansion in West German industry. Relative to the fluctuating volume of expansionary investment, the German economy shows a rationalization bias.

Table 9–1. Industrial Investment Targeting in West Germany, 1956–1978, According to Annual Census of Investment Plans.

The Rationalization Bias

	Type of Investment	
Year	*Expansionary*	*Rationalizing*
1956		
57	6.0	7.3
58	5.6	7.6
59	4.9	8.7
	6.0	9.2
1960	8.6	11.0
61	9.8	12.0
62	8.7	13.2
63	7.8	12.1
64	8.5	12.4
1965	10.3	12.2
66	9.6	13.0
67	6.1	13.2
68	7.7	11.0
69	15.6	11.2
1970	18.8	12.1
71	16.6	12.1
72	12.0	13.8
73	12.5	12.8
74	9.5	12.6
1975	6.2	13.0
76	6.4	11.9
77	6.9	11.9
78	7.5	12.2 (estimate)

Data Source: Ifo-Schnelldienst 5/77.

The dominance of rationalization over expansion since about 1960 is also reflected in data on industrial innovations, mostly from the United States, for the period 1952–1973. These data, collected by a group of innovations researchers for the Science Indicator Project, show a rationalization bias in the number of mainly rationalizing innovations as compared to the number of mainly expansionary innovations (see Table 9–2). Again, as in the case of the West German investment data, we observe a fluctuating number of expansionary innovations between 1952 and 1973 and a growing number of rationalizing innovations since around 1960. In addition, the fluctuations in the number of expansionary innovations are associated with business cycle fluctuations, whereas the number of rationalizing

Table 9–2. Industrial Innovations in Western Countries (mainly U.S.), 1952–1973, According to Sampling by Experts.

The Rationalization Bias

Period	Type of Innovation: Expansionary	Type of Innovation: Rationalizing
1952-1954	29	36
1955-1959	8	33
1960-1964	18	44
1965-1969	12	55
1970-1973	17	70
1952-1973	84	238

Data Source: NSF Science Indicators Report.

innovations increases steadily without a sign of coincidence with upswings and downswings in the business cycles.

However, the rationalization bias is intimately linked to the recessive type of structural change in the Western economies since 1960. For example, the allocation of labor (measured in worker's hours in industry per year) and of investment capital (measured in terms of purchases in equipment) shows remarkable signs of distortion. In most industrial nations, capital input increased from 1950 until about the time of the oil crisis (which fell into a time of slump that might in actual fact have come about without the oil shock), whereas labor hour input started to dwindle many years earlier. It was around 1964 in Dutch industry (Figure 9–1) and in British industry (Figure 9–2); somewhat earlier, around 1960, in the industry of West Germany (Figure 9–3); and somewhat later, around 1968, in Japan (Figure 9–4) and in U.S. industry (Figure 9–5). Actually, these deviations from a distinct pattern are amazingly small. Relative scarcity of labor (West Germany) or an abundance of it (Japan, United States), coupled with exports of civilian (Japan) and military (United States) services and commodities on a large scale, can sufficiently explain why some countries came earlier and others later into the grip of the rationalization bias.

Since about 1960, the rationalization bias in industrial investment was labor saving; surely, it was saving labor hours at the expense of more capital input. Between 1970 and 1976, it succeeded in displacing workers even more quickly from employment than it succeeded in shortening the work week of the employed. For example, in West German industry, the following figures illustrate this point.

Figure 9–1. Factor Allocation in Dutch Industry 1950–1976.

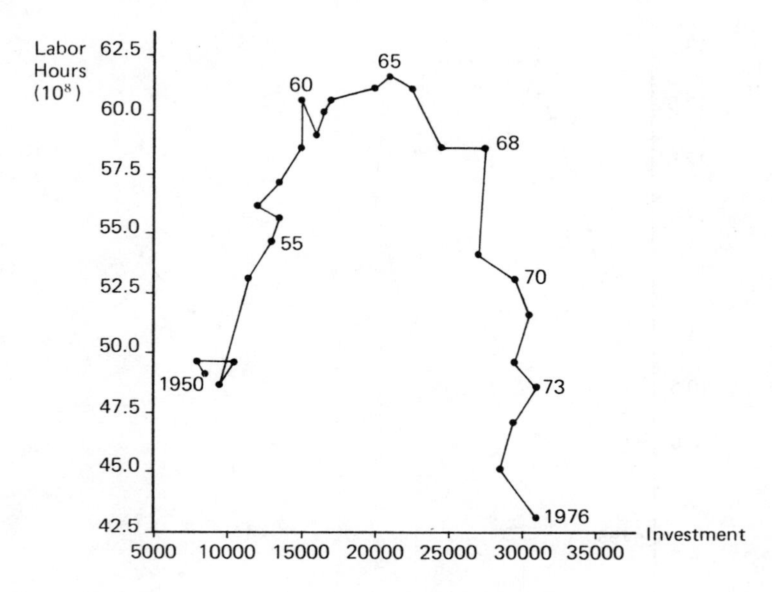

Figure 9–2. Factor Allocation in British Industry 1956–1975.

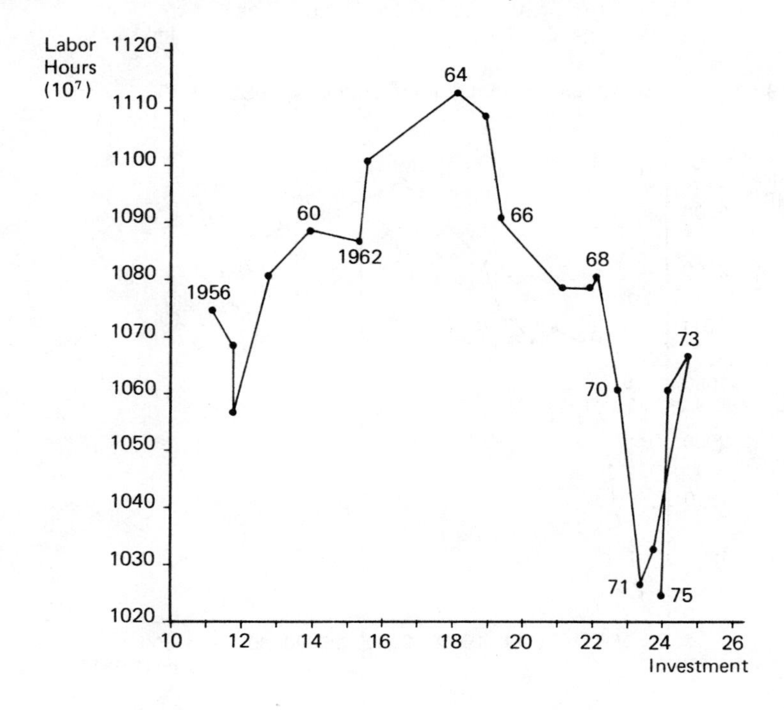

Figure 9–3. Factor Allocation in West German Industry 1950–1977.

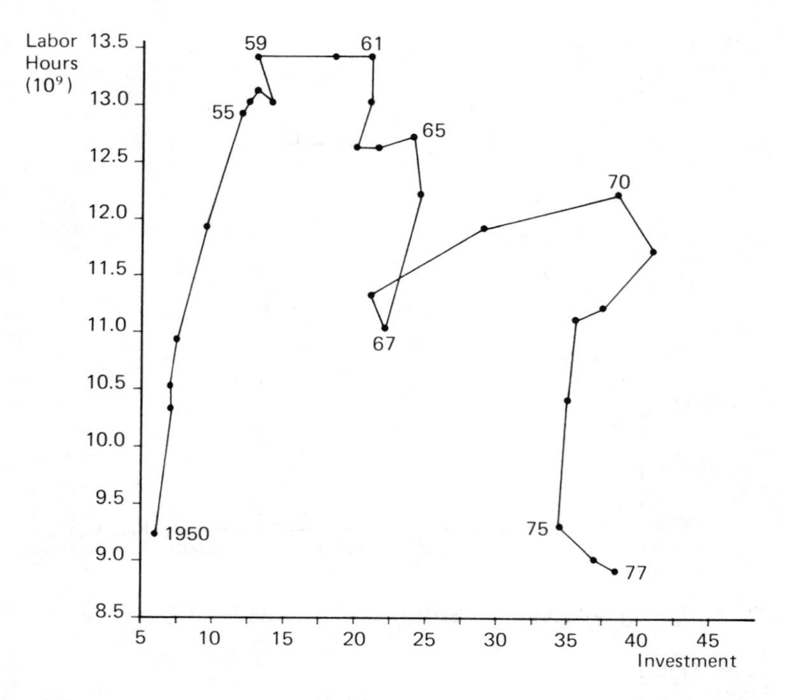

Figure 9–4. Factor Allocation in Japanese Industry 1952–1975.

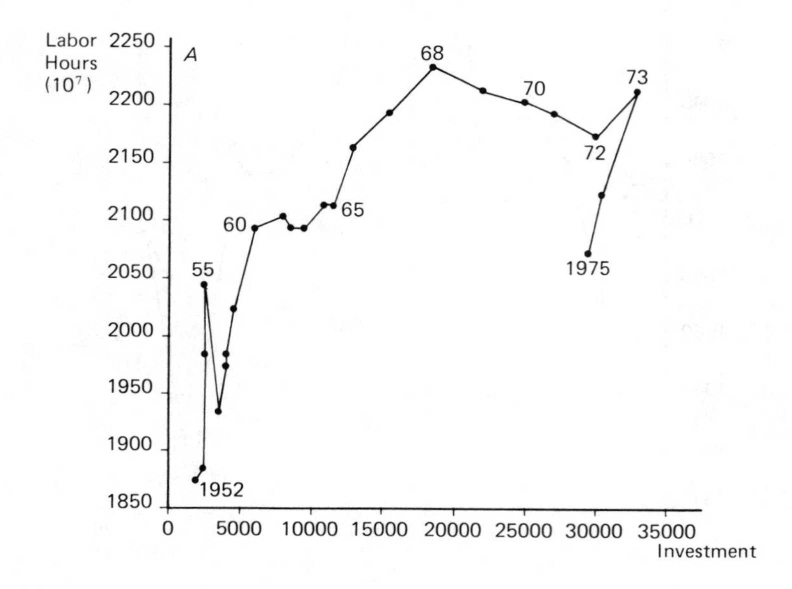

Figure 9–5. Factor Allocation in U.S. Manufacturing Industry 1950–1975.

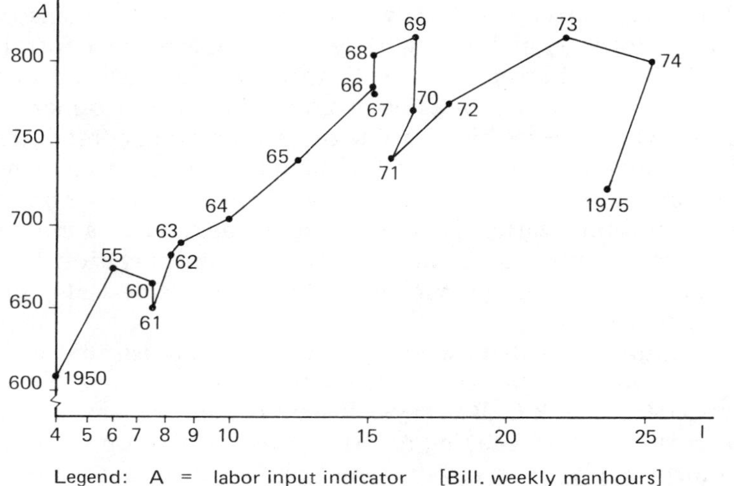

Legend: A = labor input indicator [Bill. weekly manhours]
I = capital input [Bill. of current dollars]

Data source: Stat. Abstracts of the USA, 1976, Series 1303 ILO-yearbooks, 58er, 66er, 76er.

Labor Squeeze in German Industry	*1970*	*1973*	*1976*
Million Workers	10.3	9.5	8.5
Working Hours per week	43.9	41.9	42.0

As a result of a strong tendency for rationalization and a weak tendency for expansion, production in industry stagnates in the 1970s, and industry can deliver the stagnant volume of production with fewer and fewer workers. Figures 9–1 to 9–5 show that the substitution of labor has brought the path of factor allocation into the South-East corner of the graphs. The industrial system is "cornered" at present. In economic history, the labor market recovered from such a distressed situation, in which most established industries had been cornered by the rationalization bias, only after leapfrog reactions had occurred in the industrial system. They were quantum jumps in traditional industries (plant size increases in order of magnitude) and quality jumps into new industries (basic innovations).

The capital market institutions, whose function it is to allocate investment funds in either type of projects, must be expected to prefer financing quantum jumps over financing quality jumps in future years no less than they have done in the past. In the eighteen years between 1948 and 1965, the 200 largest corporations in the United States acquired 2,692 firms (worth 21.5 billion dollars); in the three years after 1965, they acquired 1,200 more firms (worth 30 billion dollars); and today they are estimated to control two-thirds of the total assets of all U.S. firms. In the past two decades, the antitrust policy has practically given way to a neo-mercantilist approach, which places higher value on international competitiveness and technological efficiency of few suppliers than on labor market stability and a welfare optimal degree of competition among many suppliers. Rationalization is going to continue, and it is most likely to continue proceeding in the direction of concentration in established industries through the employment of large-scale, science-based, high technology.

In fact, many scientists worry today about the use that will most probably be made of the knowledge they generate; among them, Harvard president Derek C. Bok says, "There is a special, selective threat to the most novel, risk-taking forms of science, because when money is in short supply, it is always tempting to concentrate support on the proven investigator tilling familiar fields."[10] Many scientists fear that the technology bias and the rationalization bias in the economy will favor those research and development projects that pave the road for hyperindustrialization.

THE CHALLENGES AND CHANCES OF THIS STALEMATE IN TECHNOLOGY

My interpretation of the situation in the industrial world is that very many people lean toward a change in their way of life and that the economy suffers from not supplying in full measure the kind of goods and services that the public at large would buy if they were offered. As always in a stalemate in technology, industry faces a *new frontier.* The *new frontier* is a grey zone within the exchange economy, separating, on one end, the territory of goods for which it is easy to collect payment, directly by market prices and fees, or indirectly by taxation and transfer payments, and on the other end, the realm of goods for which it is difficult to collect payment, now or in the future, in terms of money or credit. This realm of value-adding activities I call the "participative economy."

Basically, the industrial economy has two different ways of adjusting to the stagnation in its fields on its side of the new frontier. It may intensify and further rationalize the industrial production, as has recently been perceived by M.I.T. president Jerome B. Wiesner. He states, "There is a growing need for sophisticated replacement technology."[11] This type of restructuring industry may mean a "hard" transition. It will force many workers to quit employment in industry and settle in other types of employment. As the private sector on one side of the new frontier has largely been occupied by industry, the slack will be provided by government expenditure for types of public demand of the usual sort—larger weapon systems, expensive infrastructure, and other more or less beneficial public works. William Wicken's proposed cure of unemployment, namely, "The improvement of English roads urged, during the existing dearth of employment for the poor," London, 1820, extends straightforward into the space age. It is common belief today that the new frontier is the terrestrial atmosphere and not the interpersonal atmosphere among people on earth. Our civilization has developed the capacity for getting rich (in terms of monetary income) from reaching out for the planets, but it has not yet developed the capacity for getting richer (in terms of satisfaction) from reaching out for fellow men.

The alternative to a hard transition into hyperindustrialization is a soft transition into a postindustrial society with a larger proportion of values being created in the "participative economy" that extends beyond the boundaries of the exchange economy. It is an ancient utopia that common welfare will be enhanced if the exchange economy is done away with and is substituted by a "participative economy" of sharing. All times know some prophet, messiah, or reformer who preached this substitution, and all times have seen these men and women being stoned to death, crucified, or being made martyrs by some other ingenious method of demonstrative monstrosity. An economy of sharing, if you think it through, also lies at the roots of utopian ideologies like communism, as it was originally perceived. Communism, having served the purpose of justifying dictatorship in many countries, has become unacceptable for freedom-loving people. On the other hand, we may use the attraction of Euro-Communism for many voters in several Western countries as an indicator for the longing of people for some change in their way of life. They seek a life that enables them to spend more time doing meaningful things together and participate in activities that, to the economist, are value-adding like favors and services given with or without the expectation of reciprocal favors and contributions.

Unfortunately, the concept of a participative economy of sharing has been discredited by historical events to such a degree that today it meets with a cultural bias against it. Hence, many traditional forms of sharing have disappeared, degenerated into rudiments of primitive or outmoded customs, or become a sign of protest. Some people frown upon experiments in ways of life that involve a certain extent of economy of sharing; others discourage it more actively. In short, the new frontier has become somewhat of a tabu in modern times. This tabu functions as a barrier to social invention and social innovation; it prevents the scientific estate from devoting resources into understanding the modus vivendi of sharing. It discourages experiments in value-adding leisure activities, and consequently it hinders the evolution of productive forms of leisure. Therefore, it is very likely that in the coming transitional years we shall not experience the emergence in sufficient numbers of basic innovations in productive, creative, and meaningful types of leisure activities. Instead we shall see the hard transition into hyperindustrialization create more demanding, intensified, and stress-burdened jobs that consume in shorter working hours the better part of the psychic energy. An extended fraction of time will be left to the dull occupations prescribed by the recreation industry and an entertainment service that provides the kind of kicks that mentally exhausted, passive consumers will certainly demand. Leisure time will be subject to structuring by professionals who make their living doing this because it is desired, and soma will become a major line of business in the Brave New World.

Of course, the two scenarios of a hard and a soft transition into the leisure economy are extremes. The challenge of this stalemate in technology is that the transition will be softened by a number of basic innovations in the underdeveloped territory lying behind the new frontier, basic innovations that introduce some modus vivendi in the realm of sharing, mutual enjoyment, and joint cultural enrichment. This stalemate in technology offers a chance of converting leisure time into more meaningful, value-adding, creative ways to live. Modern technology can help making it possible.

SUMMARY

The current situation in the industrial West suggests that many people are leaning toward a change in their way of life. On the one hand, they are disenchanted with their traditional consumption patterns and go slow on buying standardized goods (stagnation). On the other hand, they cannot articulate their unfulfilled needs. If they are not offered something that fits their shifting needs (lack of innovation), their desires remain but the economy does not produce according to existing capacities.

With such discrepancies between the need structure and the supply structure, many of the outmoded goods and services cannot be sold, and capital and labor quit these areas of technology. Due to the unfulfilled needs, labor unemployment, and capital idleness, the industrial economies become structurally ready for another rush of basic innovations. Western civilization has met with a New Frontier, and this frontier can be overcome with radical changes in technology, social organization, and political practice, which is nothing short of another Industrial Revolution.

I have shown in this book that several past industrial revolutions have been brought about by spurts of basic innovations. There are many early indicators for another rush of basic innovations in the 1980s. The question, therefore, is: In which direction will the industrial evolution go? Will it be more in the direction of technological innovation in large-scale operations ("hard" transition), or will it be more in a direction of social innovation in small-scale, participative operations ("soft" transition)?

According to historical experience (the seniority principle of innovative change), we know that tradition guides the transition. Ongoing tendencies (the technology bias, the rationalization bias) and powerful traditions push the transition in the "hard" direction, toward hyperindustrialization. The odds are not in favor of a "soft" transition but in favor of a move toward hyperindustrialization.

In the hard direction, the New Frontier is the borderline of the terrestrial atmosphere, and technological basic innovations may extend this *natural limit* and tap new sources of energy, raw materials, and expandable living space. Our civilization has developed the capacity for reaching out for the planets. But we have not advanced very far in the soft direction. We spend a diminishing fraction of our physical and mental energies on reaching out for our fellow men. Many of our facilities to share in noncommercialized activities have degenerated and have been replaced by activities for which it is easy for us or for the state to collect payment. The New Frontier, which is necessary and worthwhile to expand with social innovations, may simply be a *cultural limit* in our interpersonal "atmosphere." Pragmatism boils down to the philosophic of resignation: "Modern technology is like Shakespeare's Kate, hard to live with—and without." This philosophy leads to a "hard" transition; it may soon make us wanting to quit the spaceship earth. For a "soft" transition, we used a philosophy of symbiosis which strengthens our ties with the natural and cultural environment on earth. "Let us, then, be up and doing" (H.W. Longfellow).

REFERENCES

1. W. Goldschmidt, *Exploring the Ways of Mankind* (New York: Holt, Rinehard, and Winston, 1966), p. 115.

2. K.W. Deutsch, et al., "Political Community and the North Atlantic Area," in *International Political Communities* (New York: 1966).

3. J.M. Keynes, *General Theory of Employment, Interest, and Money* (New York: Harcourt, Bruce, 1935), p. 379.

4. J. Robinson, *Economic Philosophy* (Garden City, N.Y.: Anchor Books, 1964), p. 134.

5. T.R. Malthus, *Principles of Political Economy* (London: John Murray, 1820), pp. 401–402.

6. S. Kuznets, *Modern Economic Growth* (New Haven: Yale University Press, 1966), p. 83.

7. Ibid., p. 80.

8. Ibid., p. 85.

9. Malthus, op. cit., p. 401.

10. *Technology Review* (July/August 1976), p. 58.

11. Ibid., p. 56.

About the Author

Gerhard Mensch is Professor of Natural Science at the University of Berlin. He studied at Bonn, Stanford and Berkeley and lectured at Tulane University and the University of Bonn.

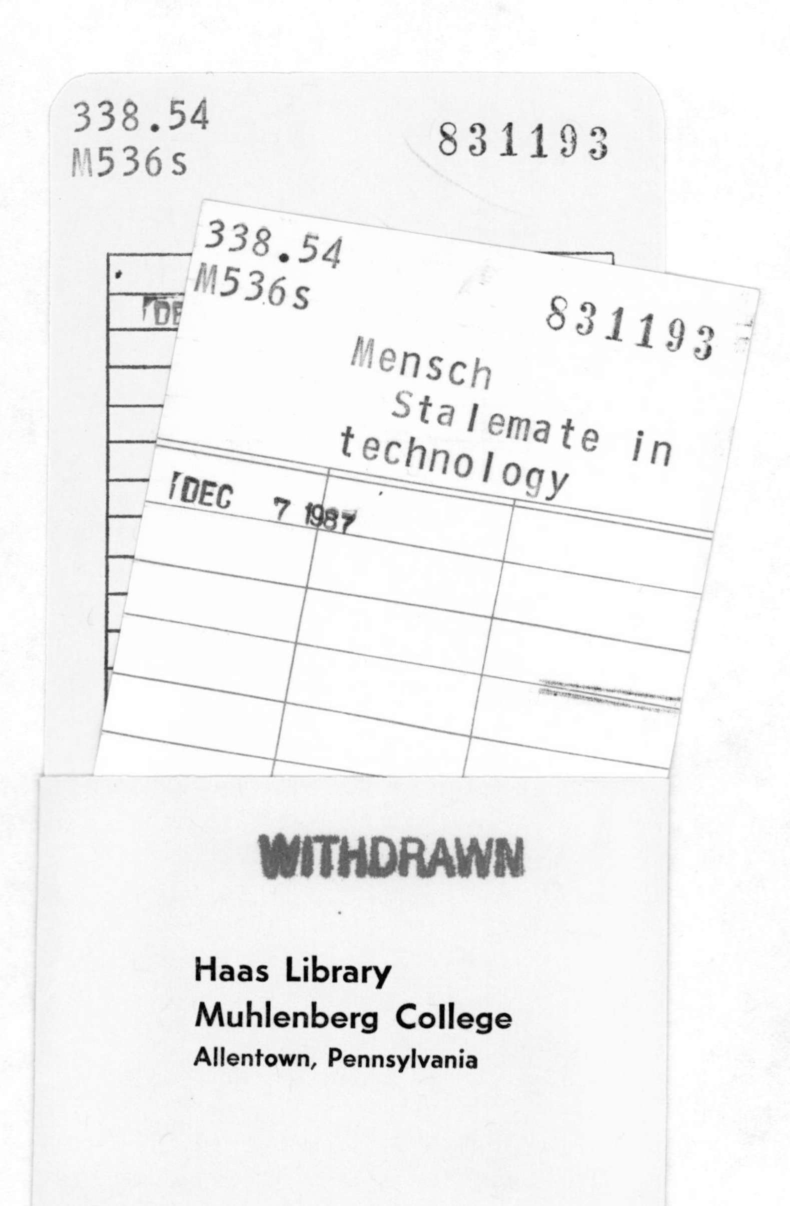
338.54
M536s
831193
338.54
M536s
831193
Mensch
Stalemate in
technology
DEC 7 1987
WITHDRAWN
Haas Library
Muhlenberg College
Allentown, Pennsylvania